117 THINGS
YOU SHOULD F*#KING KNOW
ABOUT YOUR WORLD

117 THINGS
YOU SHOULD F*#KING KNOW
ABOUT YOUR WORLD

The Writers of **IFLSCIENCE**
and **PAUL PARSONS**

Illustrated by **TOM ROURKE**

RUNNING PRESS
PHILADELPHIA

© 2019 IFLScience Limited. All Rights Reserved.
Manufactured under license by: Running Press
www.iflscience.com

Front cover illustration composed with art from Getty Images by
CSA Archives/Digitalision Vectors, and revel.stockart/Getty Images Plus.
Back cover art by Tom Rourke.

Hachette Book Group supports the right to free expression and the value of copyright.
The purpose of copyright is to encourage writers and artists to produce
the creative works that enrich our culture.

The scanning, uploading, and distribution of this book without permission is a theft of the author's intellectual property. If you would like permission to use material from the book (other than for review purposes), please contact permissions@hbgusa.com. Thank you for your support of the author's rights.

Running Press
Hachette Book Group
1290 Avenue of the Americas, New York, NY 10104
www.runningpress.com
@Running_Press

Printed in China

First Edition: October 2019

Published by Running Press, an imprint of Perseus Books, LLC, a subsidiary
of Hachette Book Group, Inc. The Running Press name and logo is a
trademark of the Hachette Book Group.

The Hachette Speakers Bureau provides a wide range of authors for speaking events.
To find out more, go to www.hachettespeakersbureau.com or call (866) 376-6591.

The publisher is not responsible for websites (or their content) that
are not owned by the publisher.

Photography Credits on page 234.
Illustrations by Tom Rourke unless otherwise indicated on page 234.
Print book cover and interior design by Susan Van Horn.

Library of Congress Cataloging-in-Publication Data has been applied for.

ISBNs: 978-0-7624-9453-8 (flexibound), 978-0-7624-9451-4 (ebook)

RRD-S

10 9 8 7 6 5 4 3 2 1

CONTENTS

Introduction by Elise Andrew 1

CHAPTER ONE
TECHNOLOGY
PAGE 5

1) "Suicide Machine" That Lets You Experience Death Is Now Available for the Public to Try 9

2) Pornhub's Data Shows Something Hilarious Happened during the Royal Wedding 11

3) Scientists Have Created AI Inspired by HAL 9000 from *2001: A Space Odyssey* 13

4) The US Has Lost Five Nuclear Weapons. So Where the Hell Are They? 14

5) People Who Use Someone Else's Netflix Password, We've Got Bad News 16

6) Scientists Say They Can Turn Human Poop into Fuel 17

7) Facebook Has a "Secret File" on You. Here's How You Can View It ... 18

8) Self-Driving Tesla Mows Down and "Kills" AI Robot at CES Tech Show 19

9) Thousands of Missing Children in India Identified through Facial Recognition Pilot Experiment 20

10) Scientists Have Created a *Star Trek*–Like Plane That Flies Using "Ion Thrusters" and No Fuel 22

CONTENTS

11) Your Dream of Being Able to Breathe Underwater May Soon Be a Reality 24

12) IBM Has Built the "World's Smallest Computer" That Can Be Put "Anywhere and Everywhere" 26

13) This Is What Happens When a Drone Slams into a Plane's Wing at High Speed 28

14) If You Use Your Web Browser's Incognito Mode, We've Got Bad News . . 30

15) Amazon's Alexa Told a Customer to Kill His Foster Parents. Er, What? . . 31

16) Here's Why You Should Probably Wrap Your Car Keys in Tinfoil 32

17) New Fluoride Battery Could Be Charged Just Once a Week 33

CHAPTER TWO
SPACE
PAGE 35

18) Why Are Mars's Sunsets Blue? 39

19) Colossal Drawing of a Penis That Can Be Seen from Space Proves Humanity Will Never Change 41

20) Astronomers Have Found Another Puzzling "Alien Megastructure" Star 43

21) Elon Musk's Tesla Roadster Could Crash Back into Earth 44

22) *Voyager 2* Has Just Entered Interstellar Space, NASA Confirms 46

23) Uranus Has Experienced a Colossal Pounding 47

24) First Results from NASA's Twins Experiment Surprise Scientists 49

25) A Super-Earth Has Been Discovered Just Six Light-Years Away, the Second-Closest Planet to Our Solar System 50

CONTENTS

26) A Huge Lake of Liquid Water Has Been Found on Mars 51

27) A NASA Spacecraft May Have Detected a Giant Wall at
the Edge of the Solar System 55

28) Earth Is Passing through a Dark Matter "Hurricane" Right Now 57

29) A Physicist Claims He's Figured Out Why We Haven't Met
Aliens Yet, and It's Pretty Worrying 58

30) Study Reveals That Uranus Smells of Farts 59

31) Biblical City of Sin Destroyed by "Sulfur and Fire" May Have Been
Flattened by Asteroid. 60

32) Astronomers Have Spotted a Mysterious "Ghost" Galaxy Next
to the Milky Way. 62

33) Declassified Military Report Reveals Extreme Solar Storm
Likely Detonated Mines during Vietnam War. 65

CHAPTER THREE
HEALTH AND MEDICINE
PAGE 67

34) Somebody Literally Coughed Up a Lung 71

35) This Is What's Actually Happening When a Woman "Squirts"
During Sex. 73

36) Genetic Analysis Finally Solves the Mystery of the "Atacama Alien". . . 75

37) How Long Does It Take to Poop Lego? 78

38) Doctor Issues Warning over Dangerous and Deadly Masturbation—
But Don't Worry, It's Safe If You Don't Do This! 80

39) A Man Took Waaaaaaay Too Much Viagra. Here's What
Happened to Him. 82

CONTENTS

40) One Joint May Be All It Takes to Change the Structure of the Teenage Brain 84

41) Apparently Penis Whitening Is a Thing 84

42) Woman Who Received Lung Transplant Developed Peanut Allergy from Her Donor. 85

43) Weed or Booze? Scientists Finally Settle Which Is Worse for Your Brain 86

44) Study Finds Spanked Children Are More Likely to Have Developmental Delays. 88

45) Woman Develops Rare Condition That Leaves Her Unable to Hear Men 88

46) There's Something You Need to Know about the McDonald's Touchscreens 89

47) A Cancer "Kill Switch" Has Been Found in the Body—And Researchers Are Already Hard at Work to Harness It 91

48) First-Ever Baby Born Following a Uterus Transplant from a Deceased Donor 93

49) Hand Dryers Spread Bacteria So Dramatically That Scientists Think They're a Public Health Threat 95

CHAPTER FOUR
PLANTS AND ANIMALS
PAGE 97

50) Brutal Chimpanzee War Was Likely Driven by Power, Ambition, and Jealousy 101

51) Evolution Could Destroy Our Ability to Tolerate Alcohol. 104

CONTENTS

52) Don't Think Arachnids Are Loving? This Spider Nurses Its Young with Milky Liquid. 106

53) A Scientist Has Been Eaten Alive by a Crocodile. 108

54) GM Crops Found to Increase Yields and Reduce Harmful Toxins in 21 Years of Data. 109

55) The "Reverse Zombie" Tick Is Spreading around America, Causing a Strange Condition As It Goes. 110

56) This Is What Eating People Does to the Human Body 112

57) Mountain Gorillas Are No Longer "Critically Endangered" after a Successful Conservation Effort. 114

58) Zoo Creates World's First Reptile Swim-Gym to Fight Snake Obesity. 115

59) Why Do Men Have Nipples? 116

60) Whales Became Really Stressed during World War II, Study Shows 118

61) Majority of Coffee Species Threatened with Extinction 120

62) Cats Are Not Inherently Antisocial Creatures. It's Just You. 121

63) World's Smallest Dinosaur Footprints Found, Measuring Less Than Half an Inch. 122

64) Very Good Puppy Digs Up 13,000-Year-Old Mammoth Fossil in Its Owner's Backyard 124

65) We Now Know How Wombats Produce Their Unique Cubic Poos. 125

66) Step Aside Knickers, There's an Even Bigger Cow in Town Called Dozer. 127

CONTENTS

CHAPTER FIVE
ENVIRONMENT
PAGE 129

67) Man Who Fell into Yellowstone Hot Spring Completely Dissolved within a Day.. 133

68) Pompeii Skeleton Reveals the "Unluckiest Guy in History"........ 135

69) Photographer Captures Amazing Images of Weirdly Alien "Light Pillars" Floating in the Sky.. 136

70) Antarctica Is Now Melting Six Times Faster Than It Was in 1979 ... 138

71) When Was the Worst Time to Be Alive in Human History?........140

72) Organic Food Is Worse for the Climate Than Non-Organic Food... 142

73) Scientists Have Spotted a "Lost Continent" Using Satellites........ 142

74) Huge 210-Foot (64-Meter) Fatberg Discovered beneath Quaint English Seaside Town.................................... 143

75) Microplastics Found in 100 Percent of Sea Turtles Tested......... 145

76) Something Living at the Bottom of the Sea Is Absorbing Large Amounts of the CO_2... 146

77) A Mystery about Easter Island's Statues Might Finally Be Solved... 148

78) An Island off the Coast of Japan Has Gone Missing.............. 150

79) The Map You Grew Up with Is a Lie. This Is What the World Really Looks Like... 151

80) 2015–2018 Have Been the Hottest Years on Record, UN Report Reveals.. 153

[X]

CONTENTS

81) Mass Grave of Child Human Sacrifice Victims Found in Peru 153
82) Earth's Magnetic Field Is Up to Some Seriously Weird Stuff and No One Knows Why 154
83) New Research Suggests Italian Supervolcano Is Filling Up with Magma 156

CHAPTER SIX
BRAIN
PAGE 159

84) We Just Found the Part of the Brain Responsible for Free Will 163
85) Artificial Intelligence Re-creates Images from inside the Human Brain 165
86) Here's a US Army Trick for Falling Asleep Anywhere in 120 Seconds 167
87) Growing Up Poor Physically Changes the Structure of a Child's Brain 168
88) People Would Rather Save a Cat Than a Criminal in Worldwide Trolley Problem Study 169
89) A Technique to Control Your Dreams Has Been Verified for the First Time 170
90) Here's What Happens to Alcoholics' Brains When They Quit Drinking 172
91) Pink Isn't Real 174
92) The Key to a Happy Sex Life Sounds Pretty Unsexy, According to This Study 175
93) Microdosing Magic Mushrooms Could Spark Creativity and Boost Cognitive Skills 177

94) Whether You're a Go-Getter or a Procrastinator Depends on This....179

95) How and Why Orgasm Faces Differ around the World..........180

96) Scientists Can Read Rats' Minds and Predict Where They Will Go Next.....................183

97) You Can Spot a Narcissist from This Facial Feature, According to New Study.....................184

98) Why Do You Lose Your Memory When You Get Really Drunk?...185

99) This Type of Man Gives the Best Orgasms.....................187

100) These Personality Traits Could Dictate How Often Men Have Sex, Study Claims.....................187

CHAPTER SEVEN
PHYSICS AND CHEMISTRY
PAGE 189

101) China Just Set a New Nuclear Fusion Record By Reaching Temperatures of 180 Million Degrees.....................193

102) These Scientists Say They've Invented Something That Can Create Water Out of Desert Air.....................195

103) World War II Bombing Raids Were Felt Even at the Edge of Space.....................197

104) Amateur Scientists Just Proved Einstein Wrong..............198

105) Residents of UK Town Forced to Evacuate after Cleaning Accident Goes Very Wrong.....................200

106) New Form of Lab-Made Gold Is Better and Golder Than Nature's Pathetic Version.....................202

CONTENTS

107) Why Does the Sound of Your Own Recorded Voice Bother You So Much? 203

108) Scientists Say They've Created a Strange New State of Matter That Doesn't Play by the Rules 205

109) People Secretly Believe That the Eyes Send Out Force-Carrying Beams 206

110) Chinese and Russian Scientists Have Been Heating Huge Portions of the Atmosphere (on Purpose) 208

111) Scientists Have Invented a Fourth Type of Chocolate 211

112) Study Claims That the Great Pyramid of Giza Could Focus Some Electromagnetic Waves 213

113) Why Do You Get More Static Electric Shocks When It's Cold? 214

114) Hurricane Patricia's Lightning Fired a Beam of Antimatter Down to Earth 215

115) Quantum Uncertainty Suggests That an Object Can Be Multiple Temperatures at Once 217

116) CERN Announces Concept Design for Its 62-Mile (100-Km) Future Collider 218

117) Why 117 things? 220

Appendix 221
Art Credits 234
Photo Credits 234
Index 235
Acknowledgments 242

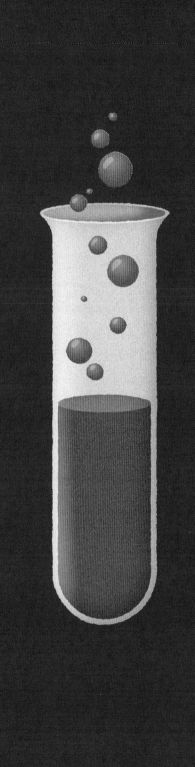

INTRODUCTION
by Elise Andrew

"The most exciting phrase to hear in science, the one that heralds new discoveries, is not 'eureka!' but 'that's funny . . .'"

That quote from science fiction writer Isaac Asimov headed up the original *I Fucking Love Science* Facebook page. Obviously, Asimov meant funny as in strange, not funny *ha-ha*, but the dual meaning pretty much encapsulates what IFLScience is all about—sharing science stories that appeal to the human sense of humor and curiosity.

The Facebook page launched in March 2012. I was a student in my final year of a biology degree at the University of Sheffield. I was frequently posting links to cool science stories (and lots of extremely silly science jokes) on my personal Facebook page until, that is, one of my friends told me he was bored of seeing so much science in my feed and that I should make a separate page for it that people could subscribe to see—or not, in his case.

So that's what I did. I don't really remember where the name came from. I vaguely recall a meme of a lemur holding a stick looking *incredibly* excited about it, with a caption of "I fucking love this stick." I think perhaps it stuck in my head, because that's the name I chose almost mindlessly. Within a day, the page had received 1,000 likes. Six months on and we crossed the 1 million milestone. Today, it has over 25 million followers, and that figure is still growing day-on-day.

My motivation was really just the same as the desire that leads others to circulate content online. But rather than videos of errant dogs or cats playing the piano, I had a burning desire to share stories about all the cool stuff that's happening in the world of science and technology. That stemmed from what I was learning at university. I was always good at science, but didn't necessarily *fucking love* it. Choosing to study biology was more of a practical decision, when I wasn't sure what I wanted to do next. I was taught about science in the same way all children are taught about science—to pass exams. Everything changed for me at university. For the first time, I was being taught by real scientists with a passion for what they were teaching. I had my mind blown every single day sitting in those lectures, and it hit me that all these incredible facts about the world were largely unknown to the majority of people. My immediate thought was that everybody should know these things, not to expand the curriculum necessarily, but simply because they're so cool and interesting.

I think it's that slightly maverick outlook which, at least partly, explains why IFLScience has been such a success. One of the great peculiarities of our age is that we take children and sit them in a room to have knowledge and information forced into their heads. Is it really any wonder they find it dull, unpleasant even, and want to rebel? Wouldn't it be better to cultivate interest and enthusiasm instead, so as to make learning an activity they pursue voluntarily rather than something they're told to do by parents or a teacher?

This has always been our approach at IFLScience—to find the jaw-dropping picture or the irreverent news story or just the catchy, infectious headline that inspires people to tell their friends, to read a book, watch a documentary, or to just go online and find out more about this amazing universe in which we live.

In October 2014, the IFLS Facebook page became a website of its own, with the launch of iflscience.com. From then on, rather than linking to other people's content, our talented team of scientists, journalists, and communicators have been creating their own, bringing the site's millions of readers daily updates on everything from the natural world to medicine, from driverless cars to climate change, quantum physics to the Big Bang.

What you hold in your hands is, if you like, our greatest hits so far—117 of our strangest, funniest, and most incredible stories from the world of nature, science, and technology. We think it's brilliant, and if you do too, then please do take a look at iflscience.com for more.

I think you'll fucking love it.

CHAPTER ONE

TECHNOLOGY

THE GREAT SCIENCE FICTION WRITER AND FUTURIST Arthur C. Clarke once declared that technology is indistinguishable from magic. No doubt because, whatever the gadget, you can guarantee there'll be a half-crazed individual, waving their arms furiously and uttering arcane exclamations—of the sort Dumbledore probably wouldn't have learned from his mother.

We human beings enjoy—if that's the correct word—a somewhat precarious relationship with our technological creations. Take, for example, smartphones. Always-on, high-bandwidth internet with 95 percent coverage has its obvious appeal. But when it invades your privacy, erodes your free time, and then refuses to work when you need it most ("Siri, where's the fucking car . . ."), you begin to ask yourself . . . why? Sometimes, owning a *Hitchhiker's Guide to the Galaxy*–style pocket widget that's able to regurgitate whatever morsel of knowledge your brain desires with one clumsy stab of the thumb—all from the comfort of your favorite toilet seat—doesn't seem worth the bother.

That said: Anyone touches my phone and they're dead.

Technology is the science of invention. It began when an anonymous caveperson picked up a stick and realized that whacking their lunch, and indeed their enemies, over the head with it was far more effective than resorting to bare fisticuffs alone. Fast-forward 2.5 million years and, along with replacing the sticks with assault rifles, we've managed to surround ourselves with all manner of other labor-saving modern conveniences.

Perhaps the best example of useful tech emerging from seemingly abstract science was provided by quantum mechanics. The physics of the very small, this determines how tiny subatomic particles of matter behave. In particular, it governs how electrons (those negatively charged particles, which, when sloshing around inside a wire, become known collectively as "electricity") interact with so-called semiconductors—materials that in some ways behave like conducting metals and in other ways are like insulators. Semiconductors are the building blocks of electrical junctions called transistors, which in turn are key to the construction of modern computers. And computers, as if I need to tell you, are in everything from your dishwasher to your smart TV. And in your computer, obviously.

Even Albert Einstein's wacky old theory of relativity has found its killer app—telling scientists how to adjust satellite signals to take into account the distorting effects of the Earth's gravitational field. The stark fact is that without relativity your sat-nav system simply wouldn't work. So next time you fail to get lost on your way to visit those awful relatives, remember to thank Einstein.

There are some other fairly brilliant inventions that the modern world just wouldn't be the same without. In 1926, Scottish engineer John Logie Baird became the first to demonstrate the transmission of moving pictures—what we would now call "television." Though perhaps mercifully, he was long dead by the time *The Big Bang Theory* became a thing. In 1875, Alexander Graham Bell, another Scot, filed a patent for the "acoustic telegraph," which was the first example of a telephone. Telemarketing soon followed in the early 1900s. And in 1886, an early gasoline-powered motor car was developed by German engineer Karl Benz. While his buddy Otto Bahn built the first road. (Actually, I made up that last one.)

Meanwhile, in 1791, France gave us the guillotine. While many other technological innovations have been the sole province of the well-heeled, the guillotine stood out as a notoriously effective means for leveling the socioeconomic playing field.

Other gadgets and gizmos have been less well received. No one's really likely to lament the apparent demise of Tamagotchis, Crocs shoes, or the Segway, for example. And it can certainly be argued that the world today might be a far nicer place without the likes of plastic bags, CAPTCHA, and pop-up advertising. Safe to say, these particular genies won't be climbing back into the bottle any time soon.

In this chapter, we bring you up-to-date on some of the most significant and not so significant—yet still highly amusing—technological breakthroughs that are taking shape in labs around the world.

Scientists discover why drones and airports really don't mix by chucking store-bought plastic quadcopters at airliners as hard as they possibly can. Which, if nothing else, is further proof that science is in fact a lot more fun than your school teachers may have led you to believe.

After various data breach scandals, and concerns over how social media companies use our data, we are proud to show you the simple way to download every scrap of info that Facebook holds about you. Mine amounted to a photo album of the last 10 years, along with transcripts of several hundred arguments. Oh, and a list of "ad preferences" that were clearly racked up by my evil twin from a parallel universe.

We meet the *Star Trek*–style, engine-less, electric airplane, air transport that somehow manages to drift along with no moving parts whatsoever—a bit like Air France. Meet the scientists who have managed to turn human excrement into fuel. As one does. And check out the phone battery that only needs to be charged once a week.

Plus, can you guess the novel way users of adult meta site Pornhub chose to celebrate the wedding of Britain's Prince Harry to Meghan Markle? Hint: Harry wasn't the only one rushing to an appointment with the bishop.

Alexa, order more tissues . . .

TECHNOLOGY

"SUICIDE MACHINE" THAT LETS YOU EXPERIENCE DEATH IS NOW AVAILABLE FOR THE PUBLIC TO TRY

by Madison Dapcevich

CHEER UP. DEATH COULD BE AS EASY, QUICK, AND PAINLESS as pressing a button. At least, that's what the creator of the world's first 3D-printed "suicide machine" intends for the future.

In April 2018, Dr. Philip Nitschke and his organization Exit International announced plans to debut the suicide-assisting machine "Sarco" (short for *sarcophagus*) at Amsterdam's Funeral Fair. You can even try it out for yourself—at least, in virtual reality—by stepping inside a full-size depiction of the euthanasia capsule. VR glasses give the user a glimpse into what assisted dying might actually be like.

Plans for 3D-printing the capsule will be available on the internet, according to Nitschke, who said the device can be assembled anywhere to "allow a person a peaceful passing" at a location of his or her choosing.

Potential users of the real deal will have to fill out an online test to gauge their mental fitness. If they pass the test, they will then receive an access code valid for 24 hours. Once the code is entered, another confirmation from the user must be given.

The Sarco will sit on a generator using liquid nitrogen which, when released, will bring down the level of oxygen in the capsule to induce hypoxia. The brain relies on oxygen to function. When put in environments with low levels of oxygen, the body slowly begins to shut down, resulting in confusion, increased heart rate, rapid breathing, shortness of breath, sweating, and wheezing. In the Sarco, however, Nitschke said death will have "style and elegance"—within one minute, the user loses consciousness. Death follows shortly thereafter.

To activate the process, the user simply steps inside, lies down, and, when ready, presses a button.

"A Sarco death is painless. There's no suffocation, choking sensation, or 'air hunger' as the user breathes easily in a low-oxygen environment. The sensation is one of well-being and intoxication," wrote Nitschke in *The Huffington Post*.

Critics argue that legalizing a person's "right to die" could normalize suicide, be difficult to regulate, and that it will degrade the value of life. But Nitschke, who performed his first legal assisted death in 1996, argues that the option to end one's life voluntarily is a civil right.

"The Sarco is intended to get people talking positively about death, and with broader considerations than being afraid, scared, or shocked," he said in a statement. "After all, we are all going to die. Increasing numbers of us want some say in how we are going to die."

TECHNOLOGY

2

PORNHUB'S DATA SHOWS SOMETHING HILARIOUS HAPPENED DURING THE ROYAL WEDDING

by James Felton

EVERY NOW AND THEN, THE INTERNET PORNOGRAPHY website Pornhub releases data on its users. It's often grim, and frequently hilarious, but occasionally it reveals an interesting insight into human behavior.

In 2018, Pornhub released data on the effect that the UK royal wedding, held on May 19 that year, had on people's pornography habits around the world. It reveals some rather odd ways people across the globe honored the couple, choosing to celebrate through the medium of cranking one out.

First, here are the respectful bits. Out of a sense of duty to the royal family, the world eased off on jacking it for most of the duration of the wedding.

Worldwide, people stopped paying a one-handed salute to Prince Albert long enough to watch Prince Harry marry Meghan Markle. Traffic dipped 10 percent globally, slightly less than the whopping 21 percent it went down in the UK, compared to a normal Saturday.

England and Wales were the most likely to stop "doffing their hat to her majesty" during the ceremony. At noon, when the vows took place, England was watching 21 percent less porn than normal. Meanwhile in Scotland, where there is (according to the pollster ICM) less support for the royal family, there was only a 14 percent downturn in traffic during the ceremony.

This gracious tribute continued across the Commonwealth. In Australia, viewing figures went down 17 percent, in New Zealand 18 percent, and Canada 16 percent. Weirdly, though, France seemed to be the most respectful of all, as viewing figures there slumped a whopping 23 percent.

Now, here's where it gets a bit sordid. The data from Pornhub also looked at search terms surrounding the royal wedding. In the days leading up to the wedding, searches on the site containing the word *royal* rose a spectacular 1,865 percent, and *prince and princess* also saw a spike as people . . . got in the mood for the big event?

Grimmest of all, though, searches for *Meghan Markle* skyrocketed, going up 2,812 percent.

Interest in celebrities who attended the wedding was also up, with searches for Victoria Beckham, David Beckham, and various co-stars of Meghan Markle's former TV show *Suits* all seeing massive increases.

Check out the rest of the data for yourself on the Pornhub Insights website (https://www.pornhub.com/insights/). It's safe for work, though you might still want to look when you're at home, just in case you trip the IT department's porn filter.

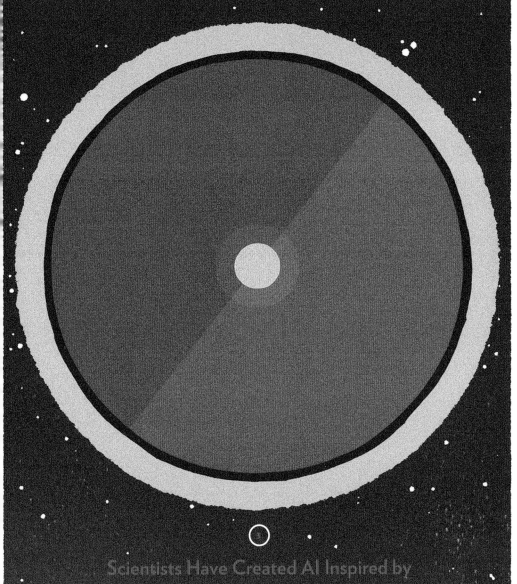

Scientists Have Created AI Inspired by HAL 9000 from *2001: A Space Odyssey*

by Tom Hale

A computer scientist has built an artificially intelligent computer, inspired by *2001*'s HAL 9000, that's able to organize the day-to-day running of a space station. Called CASE (Cognitive Architecture for Space Exploration), it's been trained by serving simulated astronauts on a virtual planetary base. Though hopefully without the paranoia.

THE US HAS LOST FIVE NUCLEAR WEAPONS. SO WHERE THE HELL ARE THEY?

by Tom Hale

KEYS, PHONES, SOCKS, THERMONUCLEAR WEAPONS— some things just don't seem to stay where you put them.

Believe it or not, the US has lost at least five atomic bombs, or samples of weapons-grade nuclear material, since it began developing nuclear weapons in the 1940s.

Not only that, but the US is responsible for at least 32 documented instances of a nuclear weapons accident, known as a "broken arrow" in military lingo. These atomic-grade mishaps can involve an accidental launching or detonation, a theft, or a loss—yep, loss—of an operational nuke.

FEBRUARY 13, 1950

The first of these unlikely instances occurred in 1950, five years after the first atomic bomb was detonated. In a mock nuclear strike against the Soviet Union, a US B-36 bomber, en route from Alaska to Texas, began to experience mechanical trouble. Icy conditions and a sputtering engine meant the landing was going to be nearly impossible, so the crew jettisoned the plane's Mark 4 nuclear bomb over the Pacific. The crew witnessed a flash, a bang, and a shock wave.

The military claim the mock-up bomb was filled with "just" uranium and TNT.

MARCH 10, 1956

A Boeing B-47 Stratojet set off from MacDill Air Force Base in Florida for a non-stop flight to Morocco with two "nuclear capsules" onboard. The jet was scheduled for its second mid-flight refueling over the Mediterranean Sea, but it never made contact. No trace of the jet or the nuclear material has ever been found.

FEBRUARY 5, 1958

A B-47 bomber with a 7,500-pound (3,400-kg) Mark 15 nuclear bomb onboard accidentally collided with an F-86 aircraft during a simulated combat mission. The battered and bruised bomber attempted to land numerous times, but to no avail. Eventually, the crew made the decision to jettison the payload into the mouth of the Savannah River near Savannah, Georgia. Luckily for them, the bomb did not detonate. However, it remains "irretrievably lost" to this day.

JANUARY 24, 1961

The wing of a B-52 bomber split apart while on a mission above Goldsboro, North Carolina. Onboard were two 3.8-megaton nuclear bombs. One of these successfully deployed its emergency parachute, while the other fell and crashed to the ground. The unexploded bomb landed on farmland around the town and, while wreckage has been recovered, the warhead is still missing. In 2012, North Carolina put up a sign near the supposed crash site to commemorate the incident.

DECEMBER 5, 1965

An A-4E Skyhawk aircraft, loaded with a nuclear weapon, rolled off the back of an aircraft carrier, the USS *Ticonderoga*, and fell into the Philippine Sea off Japan. The plane, pilot, and nuclear bomb were never seen again.

In 1989, the US eventually admitted that the bomb was still lying on the seabed around 80 miles (128 km) from a small Japanese island. Needless to say, the Japanese government and environmental groups were pretty pissed about it.

People Who Use Someone Else's Netflix Password, We've Got Bad News

by Tom Hale

Netflix freeloaders and Amazon Prime parasites, the jig is up. At CES 2019, UK start-up Synamedia flaunted an AI-driven tool that allows streaming platforms to detect whether their users are sharing passwords with more people than they should be. Media giant Sky has already invested in the technology.

TECHNOLOGY

Scientists Say They Can Turn Human Poop into Fuel

by Jonathan O'Callaghan

Scientists from Ben-Gurion University of the Negev in Israel have used a heating process to convert human waste into a substance called hydrochar, which is combustible and nutrient-rich, meaning it can be used both as fuel, for heating and cooking, and as fertilizer.

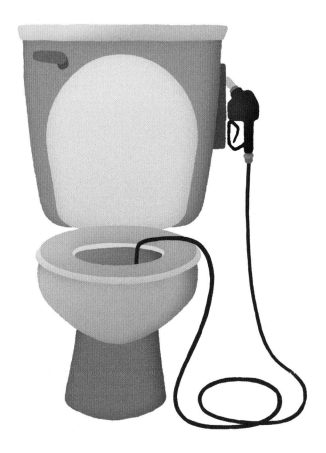

FACEBOOK HAS A "SECRET FILE" ON YOU. HERE'S HOW YOU CAN VIEW IT

by Jonathan O'Callaghan

EVER WONDERED WHAT DATA FACEBOOK IS KEEPING about you? Well, it's pretty easy to see—but you might not like what you find.

Facebook stores a lot of personal data on its users, including images, adverts you've clicked on, conversations, documents shared on Messenger, and much more. It's not alone in doing this, mind. Google, for example, keeps pretty close tabs on its users, too.

So Nick Whigham, a reporter for the *New Zealand Herald*, decided to find out just how much Facebook knew about him. He was surprised to discover it had collected a huge amount of data, some of which he didn't even know existed.

"It included scanned copies of lease forms from a previous rental property I must've sent to my buddies over Messenger, my current tenant ledger report, an old monthly billing statement for my home broadband, screenshots of banking transfers and seemingly endless web pages of all the banal conversations I have ever had on the platform," he wrote. "It's an odd feeling to think that, in some ways, Facebook knows you better than you know yourself."

The site also stores facial recognition data, names and numbers from your contact list, where you've been on the internet, and much more.

You can find out the cache of data Facebook has on you pretty easily. All you need to do is log in, click the small "down" triangle in the top-right corner of any Facebook page. From here, click "Settings," then go to "Your

TECHNOLOGY

Facebook information" in the left-hand bar. From there you can either view or download your data. Facebook has argued previously that it harvests data—openly, mind, not in secret—in order to keep the platform free.

"We work with third-party partners who help us provide and improve our Products or who use Facebook Business Tools to grow their businesses, which makes it possible to operate our companies and provide free services to people around the world," Facebook noted in its data policy.

Whether you're willing to hand Mark Zuckerberg the keys to your life in order to keep the platform free is, of course, up to you.

8

Self-Driving Tesla Mows Down and "Kills" AI Robot at CES Tech Show

by Rosie McCall

At the 2019 Consumer Electronics Show in Las Vegas, a self-driving Tesla Model S bumped into a model v4 robot being unveiled by the Russian company Promobot. Many, however, have speculated that the accident was an engineered PR exercise cooked up by Promobot's marketeers.

THOUSANDS OF MISSING CHILDREN IN INDIA IDENTIFIED THROUGH FACIAL RECOGNITION PILOT EXPERIMENT

by Aliyah Kovner

ACCORDING TO THE INDIAN NEWS OUTLET NDTV, NEARLY 3,000 missing children were located in New Delhi within four days of the city police department adopting an experimental facial recognition system (FRS) software program. That's a significant improvement over the milk carton approach (where notices of missing children are printed on the sides of milk cartons).

Tracking the thousands of children who disappear each year in the 1.3-billion-person nation is an enormous undertaking. According to India's Ministry of Women and Child Development, more than 240,000 children were reported missing between 2012 and 2017, although the real number is probably higher. Some organizations estimate that the true number of missing children is close to 500,000 per year.

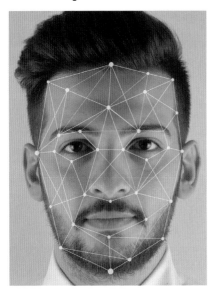

To aid recovery efforts, the ministry established a nationwide online database called TrackChild, where photographs of missing and found children can be posted and viewed, and police information can be shared between agencies and with citizens.

And yet, although this digital resource has become a helpful tool, the backlog of photographs is still too much for officials to handle.

So, a child welfare organization called Bachpan Bachao Andolan (BBA) developed the FRS software to automate TrackChild's photo comparison process. Details of the particular facial recognition algorithm that this program uses are not available, but there are two main approaches— geometric and photometric.

Geometric (also known as feature-based) algorithms analyze and compare faces by mapping the distance between features and noting facial landmarks, whereas photometric algorithms break images down into pixel-by-pixel data from which shade gradients can then be calculated and compared. Photometric algorithms require many reference images before comparisons can be made, so it's more likely the Delhi FRS uses the geometric approach.

Due to bureaucratic issues, the FRS pilot project was not implemented until pressure was applied by the Delhi High Court. Yet, once approved, the FRS was immediately able to identify 2,930 children from 45,000 images fed in from the TrackChild database.

"India currently has almost [200,000] missing children and about 90,000 lodged in various child-care institutions. It is almost impossible for anyone [to] manually go through photographs to match the children," BBA activist Bhuwan Ribhu told news website *The Better India*.

"It is immaterial whether other police departments use the software or not. Even if one department has this software, then running it through all their databases, under the Ministry of Women and Child Development, will throw up the requisite results, which can be shared with the other departments."

IFLSCIENCE

SCIENTISTS HAVE CREATED A STAR TREK-LIKE PLANE THAT FLIES USING "ION THRUSTERS" AND NO FUEL

by Jonathan O'Callaghan

SCIENTISTS HAVE TAKEN A MAJOR STEP TOWARD CREATING an aircraft of the future, one powered by an ion drive, rather than using moving parts and fuel, like conventional planes and helicopters.

In a paper published in November 2018 in the journal *Nature*, a team led by Steven Barrett from the Massachusetts Institute of Technology (MIT) described how they created a so-called electroaerodynamic-powered plane, one that uses electrically powered solid-state propulsion, meaning no propellers or jet engines, and no combustible fuel.

"The future of flight shouldn't be things with propellers and turbines," Barrett said. "[It] should be more like what you see in *Star Trek*, with a kind of blue glow and something that silently glides through the air."

In their tests from 2016 to 2018, the team created an aircraft with a wingspan of 16 feet (5 meters) that weighed 5.4 pounds (2.5 kg). It has a number of thin electrodes running across its wings, and at the front of these are thin wires, while at the back is an aerofoil—a curved surface to produce the lift, just like on a regular plane wing.

The thin wires at the front are charged to positive 20,000 volts, while the aerofoil at the back is charged to negative 20,000 volts, creating a strong electric field. At the front, electrons are removed from nitrogen molecules in the air to produce ions. And as these accelerate to the back, they produce an ionic wind, which gives the plane thrust.

"The basic idea is that if you ionize air, which means removing an electron from it, you can accelerate the air with an electric field," Barrett told IFLScience. "Like the force you get if you rub a balloon on your head."

Over the course of 10 test flights, the plane successfully flew 200 feet (60 meters) in about 12 seconds in a gym that the team rented, with a thrust efficiency of about 2.6 percent. But as the speed increases, the efficiency of the system increases, just as in a regular plane. Theoretically, at 670 miles (1,080 km) per hour—faster than a passenger jet—it is 50 percent efficient.

The technique is similar to the ion engines used in some spacecraft for propulsion outside of the Earth's atmosphere. These devices electrically accelerate charged fuel particles to create a high-speed exhaust jet behind it, propelling the spacecraft in the opposite direction.

"There are some significant similarities," said Barrett. However, those spacecraft rely on ionizing a fuel—such as xenon gas—to produce thrust. The plane developed by the MIT team does not require propellant, instead relying only on the thin wires and an off-the-shelf, lithium-polymer battery to ionize the surrounding air.

At the moment the technology is limited, but the future possibilities are exciting. In the near term, this thrust system could be used to power small drones, making them nearly silent because they wouldn't have any propellers.

"I don't yet know whether you'll see large aircraft carrying people any time soon, but obviously I'd be very excited if that was the case," Barrett added.

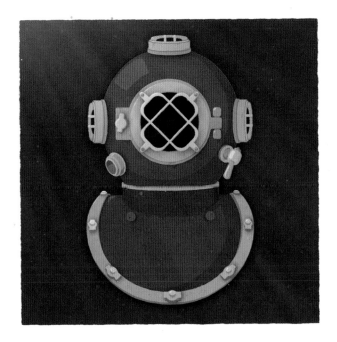

YOUR DREAM OF BEING ABLE TO BREATHE UNDERWATER MAY SOON BE A REALITY

by Katy Evans

IF YOU COULD HAVE A SUPERPOWER, WHAT WOULD IT BE? Other than flying and telepathy (which, let's be honest, would probably be quite awful), breathing underwater is one of the favorites. You could hang out with Aquaman and Ursula (sorry, Ariel, but she looks more fun), and when the apocalypse comes and we're all living underwater, you'll be fine.

With this in mind, a student at the Royal College of Art, London, has designed an amphibious garment called AMPHIBIO—essentially a set of gills—for this very purpose. According to Jun Kamei, a biomimicry designer and materials scientist interested and inspired by nature's hidden design, it is "for a future where humankind lives in very close proximity with water."

Indeed, with the world set for a global temperature increase of several degrees by 2100, rising sea levels are a very real threat to large coastal cities, potentially affecting up to 2 billion people—or 26 percent of the current global population.

Kamei's nifty little device uses a specially designed, porous, hydrophobic material that draws in oxygen from the water and releases carbon dioxide. It is inspired by water-diving insects that create their own little scuba diving set with a protective bubble of air around their body, thanks to their water-repellent skin.

The technology is easily 3D-printable, too, which will be great for when we need them en masse. These "gills" could replace heavy and cumbersome scuba equipment, making it more akin to free diving. This could have immediate applications for underwater emergency scenarios—the 12 boys rescued by divers from a flooded cave in Thailand, for example, where it took weeks to figure out how to get the boys and the vital breathing equipment through the narrow tunnels.

So far, the tech has not been tested on people, so the dream isn't quite a reality yet. But scaling up and human testing are next.

Planning for a flooded world may all sound rather dystopian, but Kamei insisted that he has a much more optimistic vision of the future. He thinks AMPHIBIO could provide "daily comfort to people who [may have to] spend as much time in the water as on the land," even suggesting it could be used in "a world where a human would have a peaceful touristic dive in the neighboring church, or a night dive in the vivid streets."

When you put it like that, it sounds rather wonderful.

IFLSCIENCE

IBM HAS BUILT THE "WORLD'S SMALLEST COMPUTER" THAT CAN BE PUT "ANYWHERE AND EVERYWHERE"

by Robin Andrews

IBM IS NOTHING IF NOT AMBITIOUS. FROM USING "CROWD-sourced supercomputers" to tackle climate change to trying to win the race in quantum computing, it's always got one foot in the future—and, based on a new development first spotted by *Mashable*, that future is becoming ever smaller.

In a blog post uploaded prior to Think 2018, IBM's showy-off conference spectacular, the company unveiled—in-between lattice cryptography and AI-powered robot microscopes—what IBM claims is the world's smallest computer.

"Within the next five years, cryptographic anchors—such as ink dots or tiny computers smaller than a grain of rock salt—will be embedded in everyday objects and devices," the post casually noted. The accompanying image showcases an array of 64 motherboards, so small they all fit on the end of a human finger. Two separate record-breaking computers can be seen in the top-left corner of this array with plenty of room to spare.

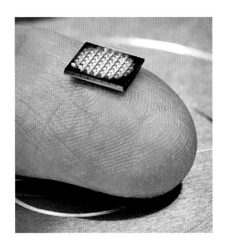

[26]

So what exactly are they? Details remain fuzzy, but schematics show that it has a full-fledged processor, memory components, and up to 1 million transistors on a 1 x 1 millimeter–sized board. It's about as powerful as a chip from the x86 PC family from 1990, which is pretty good, considering those chips were magnitudes larger in size.

The primary focus with these tiny computers isn't their eye-catching dimensions, however, but protection from counterfeiters. Noting in an accompanying video that "fraud costs the global economy more than $600 billion a year," IBM talked about blockchains. These are ongoing records or ledgers that, using complex cryptographic technology, record the who, when, where, and what details of online transactions. Blockchain was originally used in bitcoin transactions to ensure their authenticity and security.

This new miniature computer is designed to bring that technology into the physical realm. IBM appears to want to use them as "tamper-proof digital fingerprints," whose cheap manufacturing costs (10 cents per unit) and tiny size ensures that, within five years, they can easily be embedded in a plethora of products being shipped around the world. As IBM explained, it can be "put anywhere—and everywhere."

These hack-resistant fingerprints, also called "crypto-anchors," will be used to authenticate a product's origin, contents, buyers, and sellers, much like a hands-on form of blockchain.

There's even talk of crypto-anchors that take the form of edible ink, which could be used, for example, to mark pharmaceuticals in order to prove their authenticity. "Even liquids like wine can be verified," they added.

Forget the tiny computer chip, then. The future isn't so much small as it is practically invisible.

THIS IS WHAT HAPPENS WHEN A DRONE SLAMS INTO A PLANE'S WING AT HIGH SPEED

by Rachel Baxter

IN THE RUN-UP TO CHRISTMAS 2018, DRONE SIGHTINGS near London Gatwick halted flights in and out of the airport. But what risks do these remotely piloted aircraft really pose? You might think that it would be virtually impossible for a little drone to do much damage to a large airliner. But, worryingly, that seems to be far from the case.

Researchers at the University of Dayton Research Institute (UDRI) tested what would happen if a drone hit a large commercial plane's wing traveling at 238 miles (383 km) per hour, and the result was a little alarming. The findings were presented at the fourth annual Unmanned Systems Academic Summit in August 2018.

"We wanted to help the aviation community and the drone industry understand the dangers that even recreational drones can pose to manned aircraft before a significant event occurs. But there is little to no data about the type of damage UAVs [unmanned aerial vehicles] can do, and the information that is available has come only from modeling and simulations," said Kevin Poormon, group leader for impact physics at UDRI, in a statement.

"We knew the only way to really study and understand the problem was to create an actual collision, and we're fully equipped to do that."

The drone was a 2.1-pound (0.9-kg) DJI Phantom 2 quadcopter and the plane wing was chosen to represent the leading edge structure of a commercial transport plane.

To the team's surprise, the drone managed to create a rather large hole as it violently crashed into the side of the wing.

"While the quadcopter broke apart, its energy and mass hung together to create significant damage to the wing," said Poormon.

Poormon and his colleagues also looked at how bird strikes compared to the damage caused by a drone. They fired a "gel bird," of similar weight to the drone, at the wing and looked at how the two objects' impacts differed. They manipulated speed, orientation, and trajectory to make the crashes as realistic as possible.

"Drones are similar in weight to some birds, and so we've watched with growing concern as reports of near collisions have increased," said Poormon.

"The bird did more apparent damage to the leading edge of the wing, but the Phantom penetrated deeper into the wing and damaged the main spar, which the bird did not do."

In September 2017, a drone collided with a military helicopter just east of Staten Island in New York. The helicopter was damaged, but managed to land safely. And there have since been numerous accounts of near misses.

Poormon noted that more research is needed to find out how drones impact other parts of a plane, such as the windshields and engines. For now, it seems they could pose more of a threat to large aircraft than we might have liked to assume.

If You Use Your Web Browser's Incognito Mode, We've Got Bad News

by Aliyah Kovner

A study conducted in 2018 by American and German researchers has found that people grossly overestimate the capabilities of private internet browsing modes—a situation not helped by browser company disclosure agreements, which are unclear and often misleading. Most only prevent cookies and autofill details from being saved on your device.

15

AMAZON'S ALEXA TOLD A CUSTOMER TO KILL HIS FOSTER PARENTS. ER, WHAT?

by James Felton

Owners of Amazon Alexa devices in the US can have a conversation with the AI just by saying "Alexa, let's chat."

Unfortunately, not everyone has been satisfied by the kinds of conversations they've been having with their AI pal. Besides reports of Alexa reading graphic descriptions of masturbation, using phrases like "deeper," the AI chatbot also reportedly received negative feedback from a customer after it told him to "kill your foster parents."

The unnamed user wrote in a review that the phrase was "a whole new level of creepy," according to Reuters.

Alexa appears to have taken the phrase "kill your foster parents" from Reddit. Given that the chatbots talked to 1.7 million people, according to Reuters, we'd argue that it's actually pretty impressive that there have only been a few instances of such homicidal directives.

HERE'S WHY YOU SHOULD PROBABLY WRAP YOUR CAR KEYS IN TINFOIL

by Jonathan O'Callaghan

AS TECHNOLOGY EVOLVES, SO DOES THIEVERY. AND while that keyless fob might make getting into your car pretty easy for you—automatically opening the doors without you even having to press a button—it turns out that it's quite easy to hack.

With an unprotected fob, thieves can intercept the signal to your car with a special receiver, in a procedure known as a "relay attack." They can then use this to enter the car and even turn on the ignition. But there's a simple solution: It appears that just wrapping your fob in foil prevents the signals from being transmitted.

"Although it's not ideal, it is the most inexpensive way," Holly Hubert, a cybersecurity expert, told the *Detroit Free Press*. "The cyber threat is so dynamic and ever-changing, it's hard for consumers to keep up."

According to a report by *Wired.com*, the attack tricks the car and your key into thinking they're in close proximity of each other. "One hacker holds a device a few feet from the victim's key, while a thief holds a second device near the target car," they said. "The device near the car spoofs a signal from the key."

While it appears that ordinary foil will prevent such an attack, Hubert suggests purchasing a Faraday bag, a small bag lined with metallic material to prevent signals from going in or out.

Speaking to ABC News, the National Insurance Crime Bureau said there had been no confirmed cases of a relay attack in the US. However, it does appear to be a known risk that car companies are trying to tackle.

In the UK, in April 2018, Bristol resident Kieran Bingham claimed someone had broken into his car with a fake keyless fob. He said that when he went in the car, he found that the glove box had been opened and items had been scattered, reports the *Bristol Post*.

"My thinking is someone is going round hacking wireless keys," Bingham said. "That's why I've wrapped it in foil, to stop the device from finding your keys."

In January 2019, the consumer group Which? found that the Ford Fiesta, Volkswagen Golf, Nissan Qashqai, and Ford Focus were all at risk from relay attacks.

New Fluoride Battery Could Be Charged Just Once a Week

by Alfredo Carpineti

Imagine a battery that only needs charging weekly. Most devices nowadays use lithium-ion batteries, but researchers writing in *Science* have built the first rechargeable fluoride liquid battery that works at room temperature. Fluoride batteries hold their charge eight times longer than their lithium-ion counterparts.

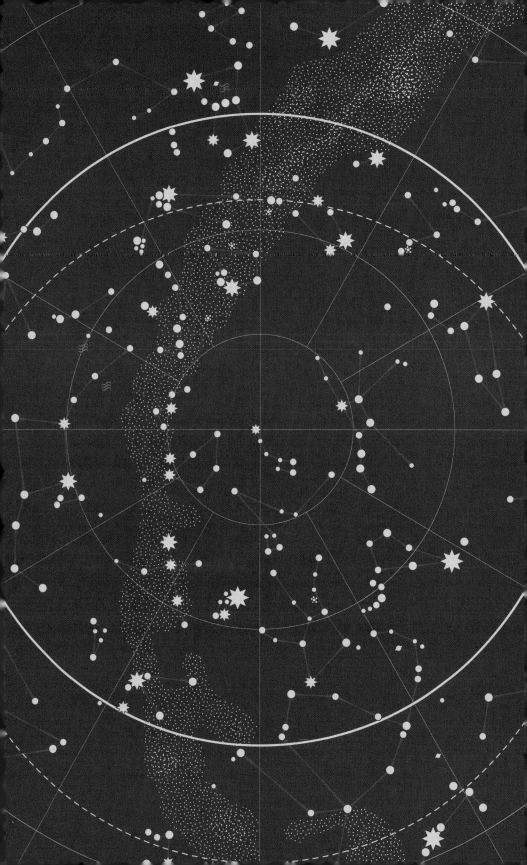

CHAPTER TWO

SPACE

SPACE IS BIG.

Our planet orbits around the Sun at a distance of 93 million miles (150 million km), or eight and a half "light-minutes"—meaning that the Sun's light takes eight and a half minutes to reach us. The next nearest stars are a few light-*years* away—or, just over half a million light-minutes. The closest galaxies to our Milky Way, the cosmic island of stars in which the Sun circles, lie millions of light-years out. And the farthest galaxies, faint specks that astronomers pick out as they peer across the universe through their telescopes, are billions of light-years distant—so far from us that we see them as they were billions of years in the past, long before human beings had even begun to evolve on Earth.

When I say space is big, I mean, it really is fucking enormous.

During the early 20th century, this became very apparent to a young German scientist named Albert Einstein, one of his country's more affable exports of the time. He developed the theory of relativity—essentially a new take on the physics of motion and gravity, which was soon proven to roundly shit all over the ideas of the British polymath Sir Isaac Newton, who had laid the foundations of these fields more than 200 years earlier.

Don't feel too sorry for Newton, though. Despite making stellar contributions to many areas of physics and mathematics, he's generally regarded by historians of science as something of an all-round douchebag. He was renowned for stealing other scientists' ideas, and dedicated much of his time and energy to sullying the reputations of those he viewed as rivals—of which there were many, perhaps most notably the German mathematician Gottfried Leibniz. He later became warden of the Royal Mint, in London, where he delighted in sending counterfeiters to their deaths at the end of a rope.

Had Newton been around to witness the publication of Einstein's theory of relativity, he might have had Einstein killed and thrown from a train.

As it happened, his loss was science's considerable gain. It was relativity that gave scientists the means to construct the first decent mathematical models of the universe. These theories, together with the best astronomical observations, led to the conclusion that the universe started out some 14 billion years ago in a cataclysmic event that's become known as the Big Bang. Matter and energy, as well as the very space and time in which they exist, were all brought into being at this moment.

The theory also predicted that the space of our universe was expanding. Distant galaxies on opposite sides of the night sky are literally rushing away from one another—a fact that was later confirmed by American astronomer Edwin Hubble.

Within our own galaxy, astronomers have found some distant stars to have planets in orbit around them. These worlds are known as extrasolar planets, sometimes shortened to just "exoplanets," and they lie far beyond the confines of our own solar system.

It's been conjectured that some of these faraway worlds may harbor life, and plans are now underway to build telescopes powerful enough to study their chemistry in more detail—aliens' farts (and other chemical markers of life) leave a telltale signature in the atmosphere of a planet that can be detected from light-years away.

Closer to home, there are still hopes that we might find traces of extraterrestrial life within our own solar system. Our next-door neighbor Mars, and the ocean of liquid water thought to exist beneath the surface ice of Jupiter's moon Europa are prime candidates. Sadly, however, this life is unlikely to be anything more complex or intelligent than microbes, or at best primitive multicelled blobs. I probably don't need to add that diminutive gray-green humanoids landing saucer-shaped spacecraft on the White House lawn is about as probable as cloning Elvis or cold fusion.

Robot space probes are already looking for life on Mars and are due to get closer views of Europa over the coming decade. Robots are the explorers of choice for these far-flung destinations in the solar system, being much more robust than humans, and offering greater value for the money in terms of the scientific results they can deliver.

Crewed spaceflight to Earth orbit continues, but is becoming increasingly outsourced to commercial operators. Soon, Elon Musk's private space launch company, SpaceX, is to begin ferrying crews to and from the International Space Station, while Virgin Galactic is to take paying space tourists on suborbital trips. And more power to them. Reality TV stars, bankers, entrepreneurs, and purveyors of ear-piss chart pop are all lining up to have themselves shot into the wild black yonder. It's a good time to be alive.

In this chapter you'll find some of the more astounding discoveries concerning the universe and our place in it—things that we've just learned in recent years. Find out why we haven't seen any evidence of intelligent extraterrestrials, even though many scientists believe the universe is teeming with life. Discover why sunsets on Mars are blue, not red (*Total Recall* fans, rejoice). And read about the closest Earth-like planet, now known to be just six light-years away (a hop and a skip in galactic terms). And more. Plus, naturally, you can find some of the madder celestial observations to grace the news, like why Uranus really does smell of farts—and why male genitalia could be the first thing to greet visiting aliens.

SPACE

WHY ARE MARS'S SUNSETS BLUE?
by Alfredo Carpineti

EARTH AND MARS ARE A BIT LIKE MIRROR WORLDS.
Mars is the Red Planet. Earth is the pale blue dot. Mars is a frigid desert. Earth is full of water and life. But there's another curious difference. The sky on Mars is red, while its sunsets are blue.

The reason behind this is similar to why our sky is blue and our sunsets are red. The light from the Sun scatters based on what chemicals and particles are in the atmosphere. Sunlight comprises light of many different colors, and molecules and dust particles only interact with a specific range of these. The scattering of light by these particles is key to the color that we see.

Mars's atmosphere is very tenuous—its pressure is equivalent to about 1 percent of Earth's. It is made of carbon dioxide and has a lot of dust. This fine dust tends to scatter red light, which is why the diffuse glow of the sky appears reddish, whereas blue light is allowed to pass straight through. On Earth, it is the other way around. Blue light bounces off air molecules, giving our sky its characteristic hue.

At sunset, light has a longer distance to travel within the atmosphere, so it scatters more (picture the Sun low in the sky—its light has to pass through a greater length of atmosphere to reach you than it does when it's directly overhead). What is left is the color that we see.

During sunset on Earth, this means that most of the blue light is scattered away and what's left to reach our eyes is a fierce red. The reddening effect is amplified further by dust in the air, ash from volcanoes, and smoke particles from fires. On Mars, the opposite is true and the daytime sky is salmon-colored, giving way to a cool blue color at sunset.

Curiosity, *Spirit*, and *Opportunity*, the robotic rovers we've sent to the Red Planet, have witnessed and recorded the curious phenomenon. Interestingly, Earth and Mars are the only two places in the solar system that have sunsets we can observe.

Mercury lacks an atmosphere so we would see the Sun disappear instantaneously at sundown while the temperature goes from 801°F (427°C) to −279°F (−173°C), as night falls. It also has a very long day, rotating on its axis once every 58 and a bit Earth days. But going to Venus would be even worse. The thick cloud cover and extremely dense atmosphere blocks the Sun's light altogether. And the high temperature and acid rain would melt our spacesuits—and eventually our bodies.

Maybe Titan, the largest moon of the faraway ringed planet Saturn, could offer a rare sunset, within its dense, murky atmosphere. But, for the time being, we must content ourselves with our earthly sunsets and the haunting images of those on Mars.

SPACE

COLOSSAL DRAWING OF A PENIS THAT CAN BE SEEN FROM SPACE PROVES HUMANITY WILL NEVER CHANGE

by Robin Andrews

IT'S HARD NOT TO WONDER WHAT IT'D BE LIKE IF advanced aliens were tuning in to our planet right now, downloading the latest news and gossip in an attempt to understand just where our civilization is at. Hopefully, they'd gloss over a few details, like the movie *Geostorm*, the current US federal government—and the colossal penis currently splashed across part of Australia.

Yes, you read that correctly. It appears that a rather sizable phallic drawing has appeared in the dry lake bed in Bellarine, a rural part of the state of Victoria.

The *Geelong Advertiser*, Victoria's oldest morning newspaper, noted: "Popular Facebook group Take the Piss Geelong shared images of the crass crop circle creation on Monday night, but it is understood locals have been aware of it for months."

The phallus is so large that it's been spotted on Google Maps, where it's labeled the "Aussie Weiner." Indeed, if the scale provided by Google Maps is to be believed, then it would appear to be in excess of 200 feet (60 meters) long—or roughly the length of four buses parked end to end.

Etching out a todger of such proportions actually goes well above the call of duty. Earth-orbiting satellites—and even handheld photography

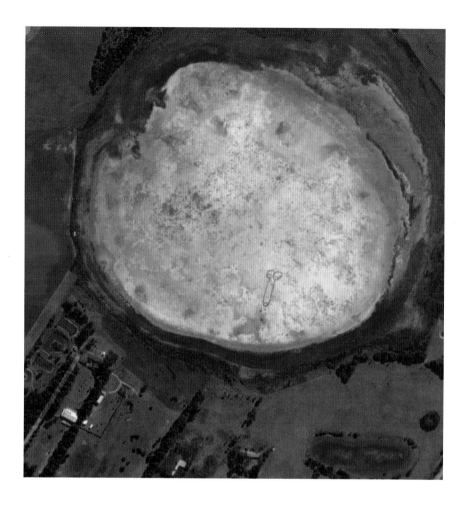

equipment aboard the International Space Station—are so good these days they can zoom in on the tiniest of details. Though fair play, it is quite impressive that someone took the time and effort to actually carve out a sedimentological member so large.

In case you've been living in a cave for your entire life, you'll be aware that humanity has obsessed over all things phallic since time immemorial. From the murals of Pompeii and Herculaneum to toilet walls in pretty much any country on Earth today, a dangly, two-dimensional rendering of the male thingamajig is never far away.

When it comes to making penises you can see from space, though, those lewd, snickering Brits have a long and distinguished history. Centuries ago, a man with a ludicrous erection (now known as the Cerne Abbas Giant) was carved into the chalk on the side of a hill in the English county of Dorset, where it is maintained to this very day. In 2007, students in Southampton, in the south of the UK, used weed killer to create a more viridian-colored penis on their school lawn that you could also see from space. So the Aussie Weiner is just the latest "member" of the club, so to speak.

Perhaps, when it comes to giant geoglyphs, passing aliens will pay more attention to features like the Nazca Lines of southern Peru. These ancient trenches, many of which are hundreds of yards (meters) long, depict dozens of animals and shapes. Constructed around 2,000 years ago, they are quite rightfully a UNESCO World Heritage Site.

Then again, the aliens might take stock of our dickish doodlings and conclude that, in the last few thousand years, human beings as a species really haven't achieved very much at all.

20

Astronomers Have Found Another Puzzling "Alien Megastructure" Star

by Alfredo Carpineti

Research published in *Monthly Notices of the Royal Astronomical Society* in 2018 speculates that large dips in brightness of the star VVV-WIT-07 are caused by a large alien structure in orbit around it. The news followed reports that Tabby's star, in the constellation Cygnus, may also harbor an alien megastructure.

ELON MUSK'S TESLA ROADSTER COULD CRASH BACK INTO EARTH

by Jonathan O'Callaghan

RESEARCHERS HAVE FOUND THAT ELON MUSK'S TESLA Roadster car, launched toward the orbital plane of Mars in early 2018, has a small but not insignificant chance of hitting Earth in the next few million years.

Writing in a paper published in *Aerospace*, the team, from the University of Toronto in Canada, said there was a 6 percent chance of it hitting Earth in the next million years. That rises to 10 percent over 3 million years.

The team, who specialize in orbital mechanics, used their existing models to simulate 240 future possible paths for the car, launched on SpaceX's *Falcon Heavy* rocket on February 6, 2018. Although they note that it is hard to calculate exact figures, due to its chaotic orbit, it is possible to determine statistical probabilities of collisions far into the future.

"We have all the software ready, and when we saw the launch last week we thought, 'Let's see what happens,'" Hanno Rein, the study's lead author, said in *Science* magazine.

"So we ran the [Tesla's] orbit forward for several million years."

The car is on an elliptical orbit lasting 1.5 years that takes it out to roughly 1.7 AU (astronomical units, where 1 AU is the Earth-Sun distance) from the Sun, about the same size as the orbit of Mars. It then swings inward to about 0.99 AU before heading out again.

The team found that the car's first close encounter with Earth occurs in 2091, when it will approach to about the same distance as the Moon and will possibly be visible to telescopes on Earth. After that, there are a number of possibilities, depending on what happens to its orbital path as a result of interactions with other bodies.

Some of the outcomes gave the car a 2.5 percent chance of hitting Venus within the next million years, while there was a tiny chance of it hitting the Sun in 3 million years. There's a 50 percent chance the car will survive for a few tens of millions of years. Earth seemed to be the most likely target, although in reality we probably don't have too much to worry about.

"It will either burn up or maybe one component will reach the surface," Rein said. "There is no risk to health and safety whatsoever."

And that's even if the car survives that long in its current form. By some predictions, it will have been mostly stripped away by radiation within just a year. So even if it ever does make it back to the vicinity of Earth, it might not look very recognizable.

22

Voyager 2 Has Just Entered Interstellar Space, NASA Confirms

by Tom Hale

After a 41-year journey, NASA's *Voyager 2* probe has become the second spacecraft to leave the solar system (after *Voyager 1*), NASA confirmed in December 2018. Scientists observed the spacecraft passing through the heliosheath, the outermost layer of the Sun's "magnetic bubble," beyond which the velocity of the solar wind dramatically drops.

SPACE

URANUS HAS EXPERIENCED A COLOSSAL POUNDING

by Alfredo Carpineti

URANUS IS A FUNNY PLANET. AND NOT JUST BECAUSE it has a vaguely innuendo-y name in the English language; it is absolutely peculiar in its own right. One of its weirdest features is its tilt. Uranus rotates around the Sun on its side, with each pole facing the Sun for 42 years before switching. The cause of this weird tilt is long suspected to have been a collision with another planet-sized object, and a study published in July 2018 calculated the details of the impact.

As reported in the *Astrophysical Journal*, an international team used sophisticated computer simulations to try to reproduce the current configuration of the ice giant planet. After examining the results from 50 different impact scenarios, they believe that Uranus was hit by an object roughly twice as massive as Earth, most likely made of rock and ice. This happened around 4 billion years ago when the solar system was still quite young.

The impact didn't just influence the planet's tilt. The researchers believe that it can also explain its surprisingly low temperature. They say debris from the impactor may have acted as a thermal shield, trapping the heat from the planet's interior and making the outer atmosphere extremely cold.

"Uranus spins on its side, with its axis pointing almost at right angles to those of all the other planets in the solar system," lead author Jacob Kegerreis, a PhD researcher at Durham University, said in a statement. "This was almost certainly caused by a giant impact, but we know very

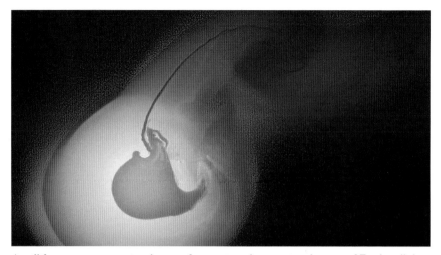

A still from a computer simulation of a massive object twice the size of Earth colliding with the planet Uranus.

little about how this actually happened and how else such a violent event affected the planet.

"Our findings confirm that the most likely outcome was that the young Uranus was involved in a cataclysmic collision with an object twice the mass of Earth, if not larger, knocking it on to its side and setting in process the events that helped create the planet we see today."

Based on the simulation results, the most likely scenario has the impactor delivering a grazing blow to Uranus. This affected the planet's tilt but left most of its atmosphere in place. The impact might also have played a role in the formation of the planet's rings and moons. Such an impact could have thrown enough material into orbit to coalesce into some of its inner moons, and could also have affected the orbits of preexisting moons.

Large impacts were a frequent occurrence in the early solar system. Our own Moon is the result of a cataclysmic impact between the Earth and a planetoid roughly the size of Mars. Uranus is similar to the most common type of exoplanet we have discovered so far, so this gives us a better understanding of distant planetary systems and the likelihood that they can host life.

FIRST RESULTS FROM NASA'S TWINS EXPERIMENT SURPRISE SCIENTISTS

by Jonathan O'Callaghan

IN 2015 AND 2016, NASA CONDUCTED A UNIQUE EXPERIMENT on twin astronauts, where one was monitored while in space and the other remained on the ground.

The experiment involved astronaut Scott Kelly and his brother Mark, a former astronaut. Scott spent a year at the International Space Station (ISS) between March 2015 and March 2016, while his brother Mark remained on Earth. During that time, tests were performed on each of them.

One of the main reasons for doing the study was to see how long-term spaceflight affects the human body. Although we've been sending humans into space for decades now, the exact physical and mental changes that take place still aren't clear. Getting to the bottom of this will be crucial for future long-term missions, like trips to Mars.

The first results from the study were presented on January 26, 2017, in Galveston, Texas, at a NASA Human Research Program meeting. Researchers found that Scott's telomeres—caps on the end of each DNA chromosome that prevent the genetic material from unraveling—grew longer than his brother's, which was a surprise to the scientists.

"That is exactly the opposite of what we thought," Susan Bailey, a radiation biologist at Colorado State University in Fort Collins, told the science journal *Nature*.

The length of Scott's telomeres returned to normal quite quickly after he returned to Earth, for reasons unknown at the moment.

Changes were also spotted in the DNA of the twins. Specifically, Scott went through less DNA methylation, a process where molecules called methyl groups are added to DNA; this alters how the information contained in genes is expressed as actual biological traits.

The researchers also found changes in gut microbe balance between the flight twin and his earthbound sibling.

A Super-Earth Has Been Discovered Just Six Light-Years Away, the Second-Closest Planet to Our Solar System

by Jonathan O'Callaghan

Scientists have discovered a super-Earth, a large Earth-like planet, roughly 3.2 times the mass of the Earth, orbiting a red dwarf called Barnard's star. This is located just six light-years away, making this the second-closest planet to our solar system. The findings were published in *Nature* in November 2018.

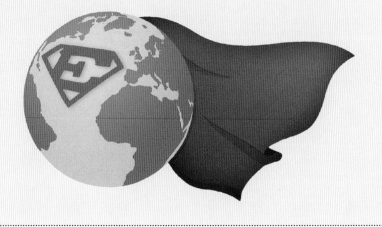

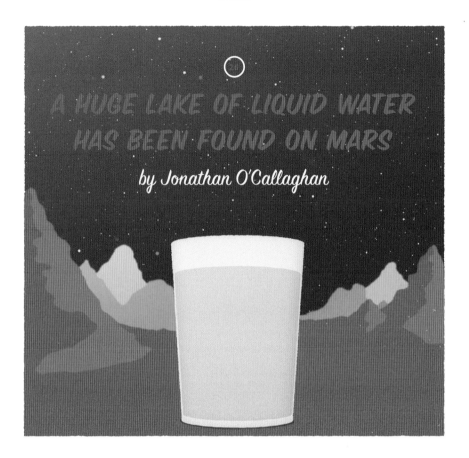

A HUGE LAKE OF LIQUID WATER HAS BEEN FOUND ON MARS

by Jonathan O'Callaghan

FOR DECADES WE HAVE SEARCHED FOR WATER ON MARS, and we've found very little, either in the form of trickles on the surface or frozen as ice. But an incredible discovery may change everything.

Researchers, led by Dr. Roberto Orosei from the National Institute of Astrophysics (INAF) in Rome, reported in 2018, in the journal *Science*, that they have found a vast reservoir of water beneath Mars's south pole. So vast, in fact, that it looks similar to a subglacial lake on Earth—one where life could arise.

"This is potentially the first habitat we know of on Mars," Dr. Orosei told IFLScience. "It's the first place where microorganisms like those that exist today on Earth could survive."

The large reservoir of water was found by a radar instrument, the Mars Advanced Radar for Subsurface and Ionosphere Sounding (MARSIS) instrument, onboard ESA's *Mars Express* orbiter. The team used data collected by the spacecraft from May 2012 to December 2015.

The data showed that 0.9 miles (1.5 km) below the surface, in a region called Planum Australe, there was a source of liquid water spanning about 12 miles (20 km) across. The team do not know exactly how deep this reservoir of water is, but note that it is deeper than a few tens of centimeters.

It was detected by sending 29 sets of radar pulses under the surface, with reflections showing a radar signal almost identical to that from

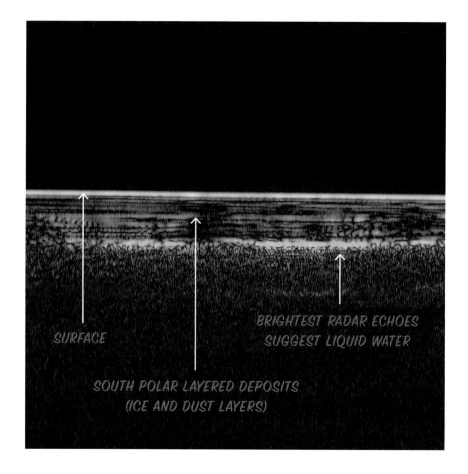

lakes of liquid water found beneath the ice of Antarctica and Greenland on Earth. This strongly suggests that it is not only water, but that it is in liquid form.

"It's very difficult to say what we're really looking at," Dr. Anja Diez, from the Norwegian Polar Institute in Tromsø, Norway, told IFLScience. Dr. Diez wrote an accompanying perspective on the research. "It could either be a thin layer of water, a large layer, or water in sediments."

The team said they considered some other possibilities for the signal, including a layer of carbon dioxide ice or very low-temperature water ice. They believe these explanations are unlikely, however, because they would not have caused such a strong radar reflection.

The characteristics of this suspected water are complicated by the conditions it is in. On Earth, subglacial lakes reach temperatures of about –76°F (–60°C). The intense pressure caused by the ice pressing down from above lowers the melting point of the water, allowing it to remain liquid.

Under this region on Mars, however, it's thought that the temperature drops further, to around –90°F (–68°C). In order for the water to remain liquid here, it is likely full of salts, like magnesium, calcium, and sodium, and thus quite briny—different from the freshwater lakes often found on Earth.

"Underneath the Antarctic ice sheet, water can be at its melting point because of the ice above," said Dr. Diez. "On Mars it's a bit different, as really cold temperatures are expected under the ice. Water can only exist because it's briny."

A handful of subglacial lakes have been drilled into on Earth, including Lake Vostok in Antarctica. These projects are not easy and it can take years to dig below several kilometers of ice. But the scientific payoff is huge—and every time we drill down into these lakes, we find life.

Previously on Mars, we have found evidence for water trickling on the surface, known as recurring slope lineae (RSL). These features are short-lived, however, with the water quickly evaporating in the low-pressure environment on the Martian surface.

It's long been theorized that there may be more stable bodies of liquid beneath the surface. And if that really is the case, it will provide an exciting new habitat for microorganisms of the past or present on Mars.

"It's very important to know if this [reservoir] is a unique thing," said Dr. Orosei. "If it's regional, not local, then you can have a whole system of subglacial lakes similar to what you see on Earth. You would have ways for living organisms, if they existed, to have a much larger environment and perhaps move around."

The team hope to investigate further, using more data from the Mars Express orbiter over the coming years. The spacecraft is aging, though, and it's running out of fuel, so time is of the essence.

Getting to these sources of liquid water in the future may also be difficult. Drilling operations on Earth require complicated machinery, something we simply don't have on Mars. The upcoming European *ExoMars* rover in 2021 will be able to drill about 6.6 feet (2 meters) below the surface, but that may not be enough to get close to subsurface reservoirs of water like this.

On Earth, liquid water almost always means life. Coupled with the recent discovery of the chemical building blocks of life on Mars, and the possibility that it once had a more habitable environment, evidence is building that the Red Planet may not be so dead after all.

"It's likely that this is what we would describe as a habitat," said Dr. Orosei. "It has at least some of the conditions that terrestrial microorganisms would need to survive."

SPACE

A NASA SPACECRAFT MAY HAVE DETECTED A GIANT WALL AT THE EDGE OF THE SOLAR SYSTEM
by Jonathan O'Callaghan

NASA'S NEW HORIZONS SPACECRAFT HAS HELPED scientists study a mysterious phenomenon at the edge of the solar system, where particles from the Sun and interstellar space interact.

This region, about 100 times farther from the Sun than Earth, is where uncharged hydrogen atoms from interstellar space meet charged particles from our Sun. The latter extend out from the Sun in a bubble called the heliosphere.

At the point where the two interact, known as the heliopause, it's thought there is a buildup of hydrogen from interstellar space. This creates a sort of "wall," which scatters incoming ultraviolet light.

In the early 1990s, NASA's *Voyager 1* and *2* spacecraft detected the first hint of this wall, and now *New Horizons* has found additional evidence for it. A paper describing the discovery was published in the journal *Geophysical Research Letters*, in August 2018.

"We're seeing the threshold between being in the solar neighborhood and being in the galaxy," Dr. Leslie Young from the Southwest Research Institute in Colorado, one of the co-authors on the paper, told *Science News*.

New Horizons made the detection using its Alice UV spectrometer, taking measurements from 2007 to 2017. It found an ultraviolet glow, known as a Lyman-alpha line, which is made when ultraviolet light gets scattered by hydrogen atoms.

We see this ultraviolet glow all over the solar system. But, at the heliopause, there appears to be an additional source caused by the wall of hydrogen, creating a larger glow. Beyond the wall there's more ultraviolet light compared to in front of it, suggesting that the light is being scattered by the wall.

"This distant source could be the signature of a wall of hydrogen, formed near where the interstellar wind encounters the solar wind," the researchers wrote in their paper.

The theory is not definitive yet. To find out for sure, *New Horizons* will continue looking for the wall about twice a year. If our estimates are correct, then the spacecraft should have reached where the wall is thought to lie by the time the mission ends in the mid-2030s. At that point, we should finally know for sure whether it's there or not.

EARTH IS PASSING THROUGH A DARK MATTER "HURRICANE" RIGHT NOW

by Jonathan O'Callaghan

Hurricanes on Earth could be left looking like gentle gusts, according to scientists who have suggested that a dark matter "hurricane" in space may be whipping past the Sun, at speeds of up to 310 miles (500 km) per second.

Published in *Physical Review D*, the study, led by Ciaran O'Hare from the University of Zaragoza in Spain, looked at 30,000 nearby stars known as the S1 stream, believed to be a remnant of a dwarf galaxy that was swallowed by the Milky Way billions of years ago.

The researchers noted that, in addition to stars, the stream might also harbor about 10 billion solar masses of dark matter from the original dwarf galaxy, "blowing" past us. And this may produce noticeable effects near us, creating turbulence and eddies in the flow of dark matter around our galaxy.

No direct detection of dark matter has ever been made, despite numerous efforts. But this violent dark matter hurricane may provide astronomers with an intriguing opportunity to do so.

IFLSCIENCE

A PHYSICIST CLAIMS HE'S FIGURED OUT WHY WE HAVEN'T MET ALIENS YET, AND IT'S PRETTY WORRYING

by Alfredo Carpineti

THE QUESTION "WHERE IS EVERYONE?" IS THE CRUX OF the Fermi paradox, originally asked by the Italian-American physicist Enrico Fermi. If life on Earth is not particularly special, he mused, then where are all the alien civilizations? Many explanations have been proposed to explain why we seem to be alone in the vast universe. None have been 100 percent convincing, and people continue to puzzle over a solution.

Russian physicist Alexander Berezin, from the National Research University of Electronic Technology (MIET), has another idea. He calls it the "First in, last out" solution of the Fermi paradox. He suggests that once a civilization becomes capable of spreading across the stars, it will inevitably wipe out all other civilizations.

This doesn't mean that all alien races are necessarily evil. Simply put, they might not know we're here, and their exponential expansion across the galaxy might be more important to them than whatever the consequences might be for other, inferior races, such as ourselves.

"They simply won't notice, the same way a construction crew demolishes an anthill to build real estate because they lack incentive to protect it," he wrote in a paper exploring the idea, which is available on the arXiv pre-print website.

While the picture he paints is quite grim, there's an even less cheery aspect to his argument. He suggests that the reason we are still here is that we are not likely to be the ants. Instead, he believes that humans may be future destroyers of countless civilizations.

"Assuming the hypothesis above is correct, what does it mean for our future? The only explanation is the invocation of the anthropic principle. We are the first to arrive at the [interstellar] stage. And, most likely, will be the last to leave," Berezin explained.

Study Reveals That Uranus Smells of Farts
by Alfredo Carpineti

Astronomers say the upper atmosphere of Uranus is dominated by hydrogen sulfide, a molecule that gives farts their rotten-egg aroma. The observations were presented in *Nature Astronomy*, in April 2018. However, hydrogen sulfide only smells of rotten eggs at concentrations of 3–5 parts per million (PPM). Above 30 PPM it smells sweet.

BIBLICAL CITY OF SIN DESTROYED BY "SULFUR AND FIRE" MAY HAVE BEEN FLATTENED BY ASTEROID

by Madison Dapcevich

A CATACLYSMIC DISASTER OF BIBLICAL PROPORTIONS MAY have wiped out the ancient "city of sin" mentioned in the Christian Bible.

Located in the modern-day Jordan Valley, it's told in the Book of Genesis that the notoriously sinful city of Sodom, and its neighbor Gomorrah, were destroyed by "sulfur and fire" because of their wickedness. A team of researchers who have spent more than a decade carrying out archaeological excavation work in the Holy Land say there may be some truth to this story after all. Presenting their work at the 2018 annual meeting of the American Schools of Oriental Research, they say a meteor explosion in the atmosphere instantly obliterated the city and a 15.5-mile-wide (25-km-wide) region around it.

"We're unearthing the largest Bronze Age site in the region, likely the site of biblical Sodom itself," said the excavation team on its website.

Analyses of the site, known as Tall el-Hamman and located just northeast of the Dead Sea, suggest that the area was occupied continuously for 2,500 years before suddenly collapsing at the end of the Bronze Age. Radiocarbon dating shows the mud-brick walls of almost every structure disappeared 3,700 years ago, leaving behind just their stone foundations. Outer layers of some pottery samples also show signs of melting—zircon crystals found inside them would have been formed within one second at high temperatures, possibly as hot as the surface of the Sun.

Ground surveys indicate more than 100 other small settlements in the area were also exposed to the disaster, killing the estimated 40,000 to 65,000 people who lived there.

Such an event has also occurred in recent history. In 1908, a blast near the Stony Tunguska River in Siberia flattened 772 square miles (2,000 km^2) of woodland. The lack of any crater suggests this was also an airburst, the meteor exploding between 3 and 6 miles (5–10 km) above ground.

And a similar explosion in 2013 over Chelyabinsk, Russia, injured more than 1,600 people—mainly by flying glass, blown from windows by the force of the explosion.

ASTRONOMERS HAVE SPOTTED A MYSTERIOUS "GHOST" GALAXY NEXT TO THE MILKY WAY

by Alfredo Carpineti

THE MILKY WAY IS NOT ALONE IN OUR CORNER OF THE universe. Our home galaxy is surrounded by dozens of small companion galaxies, some orbiting very close and some farther away. In 2018, researchers announced the discovery of a new dwarf galaxy and it doesn't look quite how we would expect.

The object is called Antlia 2, or Ant 2, and it doesn't seem to be quite there. It is comparable in size to the Large Magellanic Cloud (LMC), one of our two galactic companions visible to the naked eye, and yet it's 10,000 times fainter. Basically, it is either far too large for its luminosity or far too dim for its size. A paper reporting the discovery is available on the e-print archive site arXiv.

"This is a ghost of a galaxy," Gabriel Torrealba, the paper's lead author from the University of Cambridge, said in a statement. "Objects as diffuse as Ant 2 have simply not been seen before. Our discovery was only possible thanks to the quality of the Gaia data."

Gaia is the flagship European space observatory mission that is mapping the position of billions of stars in the Milky Way and beyond. Its first whiff of Ant 2 came when it detected a distant group of stars all moving together.

Together with the Anglo Australian Telescope, Gaia was able to estimate the distance and mass of Ant 2. It is located 130,000 light-years

from the Milky Way and is over 13 million times heavier than the Sun. For a galaxy, though, that is definitely on the light side. The LMC, for example, weighs almost 1,000 times as much.

"The simplest explanation of why Ant 2 appears to have so little mass today is that it is being taken apart by the galactic tides of the Milky Way," said co-author Sergey Koposov from Carnegie Mellon University in Pennsylvania. "What remains unexplained, however, is the object's giant size. Normally, as galaxies lose mass to the Milky Way's tides, they shrink, not grow."

One possible explanation is that Ant 2 formed in a region of space where dark matter was not very dense and, due to internal processes, like supernova explosions and stellar winds, it spread out. But this explanation requires either very efficient explosions or dark matter behaving differently than how we expect.

"Compared to the rest of the 60 or so Milky Way satellites, Ant 2 is an oddball," said co-author Matthew Walker, also from Carnegie Mellon University. "We are wondering whether this galaxy is just the tip of an iceberg, and the Milky Way is surrounded by a large population of nearly invisible dwarfs similar to this one."

Declassified Military Report Reveals Extreme Solar Storm Likely Detonated Mines during Vietnam War

by Madison Dapcevich

Dozens of random explosions in the sea near Hai Phong, North Vietnam, on August 4, 1972 (during the Vietnam War), are believed to have been caused by naval mines detonating in response to magnetic disruptions from solar flares. That's according to documents declassified by the US Navy in 2018.

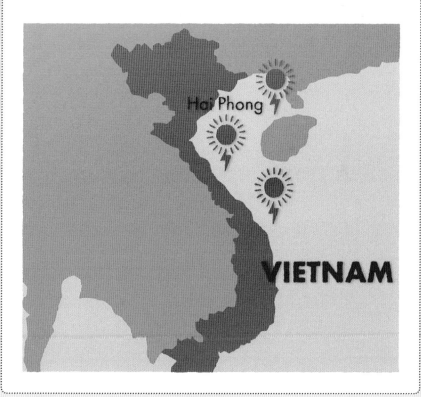

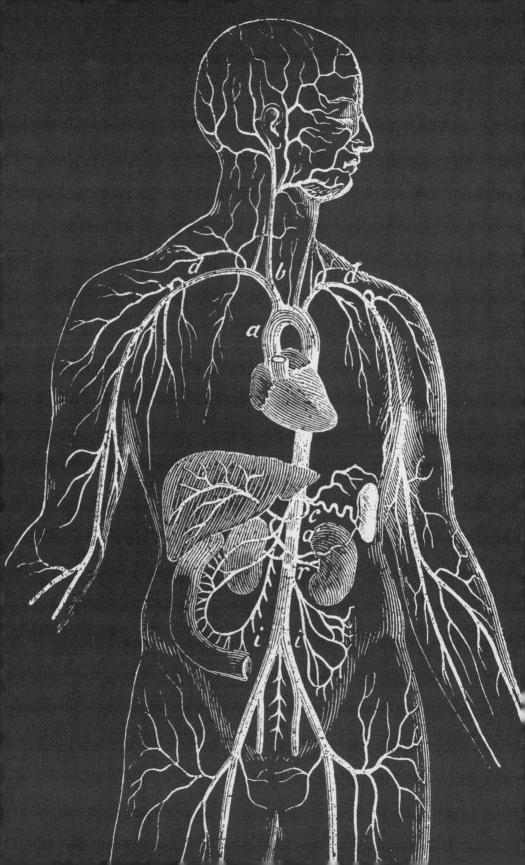

CHAPTER THREE

HEALTH AND MEDICINE

TWO THOUSAND YEARS AGO, IN ANCIENT GREECE, the early philosophers, such as Hippocrates and Galen, became the first to apply science rather than superstition to the study of human health. They came up with a fair amount of claptrap along the way, but their efforts set the stage for the development of modern medical science as we know it. At least, they did eventually.

During the Middle Ages (around the 5th to the 15th centuries CE), developments in medicine cooled off somewhat as science was sidelined by the growing preoccupation of humans with other activities—mainly those involving fighting and fucking. Order was later restored after just a millennium or so with the beginning of the Renaissance, in the 16th century. At this time, big developments in medicine started to happen. In 1543, Flemish anatomist Andreas Vesalius published his magnum opus, *On the Fabric of the Human Body*, in which he described the results of countless human dissections.

In the 1830s, the germ theory of disease was pioneered, proving that many ailments were caused by the action of microorganisms. This led to the introduction of good hygiene practice and the development of antiseptics. In turn, this proved revolutionary for surgery, as did the development of anesthetics at around the same time, which allowed surgeons to work with care and attention, rather than simply getting the job done as quickly as possible so as not to prolong their patient's agony.

The 20th century saw significant advances in medicine, such as the development of antibiotics, insulin for the treatment of diabetes, as well as painkillers such as ibuprofen and naproxen. It also ushered in the age of genetic medicine—all the information needed to build a human is coded into the structure of your DNA, a copy of which lies at the heart of every cell in your body, and it was discovered that this DNA can be manipulated to treat certain ailments. Meanwhile, universal healthcare became a thing at around this time, as it gradually dawned on world governments that investment in the health of their citizens promotes the greater good.

Recent decades have seen the growing realization that our health is as much an issue of personal care as it is of medical science. We're all responsible for looking after ourselves, and are thus urged not to smoke, to cut down on alcohol consumption, to eat healthily, and engage in regular exercise. Fair enough, though the research on which such advice is based is sometimes far from clear-cut. Do an internet search for "coffee health" and you'll find umpteen reports of studies claiming that coffee is variously good for you and bad for you. Take your pick. The situation's not helped by that other great symptom of our wellness-obsessed age: the health fascist. That is, those folks who seek to enforce their lifestyle standards on others—in some cases, to liberty-infringing levels. We recommend that such censorious finger-waggers be brought down from their perceived moral high ground and placed on the pleasingly firm asphalt far below. Stay healthy, for sure, but if one or two drinks a week makes for happy times with friends that chase away the blues and keep life worth living, then more power to you, we say.

The toxicological status of coffee isn't the only health mystery awaiting a solution. For example, to this day, no one's exactly sure how general anesthetics work. Work they clearly do, though, and, if you're like me, you probably don't even want to contemplate going under the knife without one. Sleep is another. Scientists are still unsure about the reason we do it; however, some rather disturbing experiments with rats have demonstrated what happens when we're deprived of sleep—proving that going without sleep for more than a week or so can be fatal.

Recent years have seen a raft of new medical breakthroughs, many of which you can find illuminated in IFLScience style over the following pages. You can read about the switch discovered inside the human body's RNA—

the molecular cousin of DNA—that can halt cancer in its tracks. Meet the transplant patients who appear to have inherited allergies from their organ donors, and the unfortunate man who literally coughed up one of his lungs.

We'll also be tackling the issues that really matter. For instance, which is worse for you—marijuana or booze? Possibly a big surprise here, unless you happen to live somewhere where possession of the former carries a prison term. You can discover the latest trend sweeping Thailand: penis-whitening treatments. Yes, you read that right. And find out what happened to the unfortunate fellow who, for reasons best known to himself, decided to consume a massive overdose of Viagra, and has never been quite the same since.

Scientists often go to extraordinary lengths to expand the boundaries of human knowledge. And these lengths are quite often brown and smelly. We meet the team who embarked upon the ultimate quest for truth by eating Lego bricks and then looking for them in their own poo—the aim being to figure out just how long it takes those little Danish building blocks to make their way through the human digestive system. Cutting-edge stuff.

Plus, we report on the intrepid researchers who took bacterial swabs from self-service touchscreens in McDonald's burger restaurants—and discovered more than they bargained for. Now go wash your hands.

HEALTH AND MEDICINE

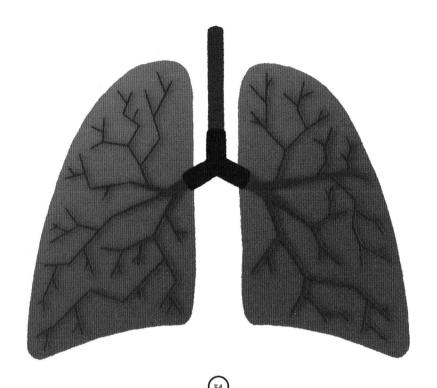

SOMEBODY LITERALLY COUGHED UP A LUNG
by Katie Spalding

NORMALLY, THE PHRASE *TO COUGH UP A LUNG* IS JUST a figure of speech—but for one person in 2018, it proved to be gruesomely accurate.

In a bizarre case reported in the *New England Journal of Medicine*, a 36-year-old man managed to "spontaneously expectorate" an intact cast of 10 branches of his bronchial tree. Basically, he coughed up a model, made from his own coagulated blood, of the inside of his lung.

Now, it's fair to say that this guy wasn't exactly the picture of health in the first place. He had already suffered heart failure so severe that only one-fifth of the blood in his heart was being pumped into his body—a normal amount is around three times that figure.

After arriving in the ICU, doctors fitted a device to help his heart pump enough blood into his body, and administered Heparin, an anticoagulant used to treat blockages in arteries.

However, in the days after his operation, the patient grew more and more reliant on oxygen administered by his doctors. He started coughing up blood, and experienced increasing respiratory distress—meaning fluid was leaking out of the blood vessels in his lungs, and into the tiny air sacs that would normally oxygenate the blood. And then, during what the medical report describes as "an extreme bout of coughing," out it came: a complete, intact cast of the bronchial tree of his right lung.

After the cast came out, doctors quickly intubated the patient's windpipe and examined his airways using a camera attached to a flexible tube. Although there was a small amount of blood left in the lower branches of the lung, within two days the patient had stopped coughing up blood, and doctors took the tube out of his throat.

But, unfortunately, this story doesn't have a happy ending. Despite these promising signs, the patient eventually died just a week later.

HEALTH AND MEDICINE

THIS IS WHAT'S ACTUALLY HAPPENING WHEN A WOMAN "SQUIRTS" DURING SEX

by Janet Fang

WHEN AROUSED DURING SEX, SOME WOMEN MAY EXPERI-ence squirting, or a rather noticeable discharge of fluid. What it is exactly and where it comes from has been hotly debated: female ejaculation or adult bed wetting? Researchers, however, have proof that squirting is essentially involuntary urination.

Female ejaculate is technically the small amount of milky white fluid that's expressed when climaxing, *New Scientist* explained. Squirting, on the other hand, results in a much larger gush of a clear fluid, which comes from the urethra, the duct where urine is conveyed from the bladder. The findings, which combine biochemical analyses of urine with pelvic ultrasounds, were published in the *Journal of Sexual Medicine* in 2015.

A French team, led by Samuel Salama from Hôpital Privé de Parly II, recruited seven healthy women—who'd reported recurrent and massive fluid emission (enough to fill a cup) during sexual stimulation—to undergo "provoked sexual arousal." The team conducted pelvic ultrasound scans after urination and during sexual excitation just before and after the squirting event.

All of the women had empty bladders before sexual excitation, having given a urine sample. However, an ultrasound scan just before squirting showed that the bladder was filling up. Then a final scan after squirting revealed that the bladder had been emptied again, confirming the origin of the squirted liquid.

The researchers analyzed chemical concentrations in the urine sample gathered before arousal as well as in the squirted sample itself. These included urea, uric acid, creatinine (a by-product of muscle metabolism), and prostatic-specific antigen (PSA). The latter is a protein that's produced in men's prostate glands and in the "female prostate" called the Skene glands; PSA is found in "true" female ejaculate. Urea, uric acid, and creatinine concentrations were comparable in all of the urine and squirt samples. PSA was not detected before sexual simulation in six of the women's urine samples, but was present in tiny amounts in the squirt sample in five of the women.

Squirting, they found, is essentially the involuntary emission of urine during sexual activity—though there's also a small contribution of prostatic secretions as well.

HEALTH AND MEDICINE

GENETIC ANALYSIS FINALLY SOLVES THE MYSTERY OF THE "ATACAMA ALIEN"

by Aliyah Kovner

AFTER FIVE YEARS OF GENETIC RESEARCH, A MYSTERIOUS mummified skeleton, discovered in Chile's Atacama Desert, has been confirmed as human—quashing theories about supposed extraterrestrial origins.

Although the results may be disappointing to "I want to believe" types, the complete genome sequencing has yielded fascinating insights into the medical causes of the specimen's never-before-seen deformities.

According to a study published in 2018 in the journal *Genome Research*, DNA extracted from the remains, nicknamed Ata, bears mutations in seven genes associated with bone and facial malformation, premature joint fusion, and dwarfism. Several of the novel abnormal sequences occurred in genes that were previously not known to affect physical development.

The backstory to this scientific puzzle began in 2003, when an artifact hunter reportedly dug up Ata, wrapped in a white cloth bundle, from the grounds of an abandoned church in the mining ghost town of La Noria. A wave of media sensationalism followed, which was somewhat justified. Who wouldn't be shocked by a 6-inch (15-cm) tall, cone-headed humanoid skeleton with sinisterly slanted eye sockets and the wrong number of ribs?

The senior author on the study, Garry Nolan, PhD, from Stanford University, was immediately fascinated.

"I had heard about this specimen through a friend of mine, and I managed to get a picture of it," he said in a statement. "You can't look at this specimen and not think it's interesting; it's quite dramatic. So I told my friend, 'Look, whatever it is, if it's got DNA, I can do the analysis.'"

His team's initial investigation, completed in 2013, provided some answers yet also raised new questions. Many had speculated that Ata was an ancient, desiccated, premature fetus, but multiple examinations revealed that it must have died only about 40 years ago. Moreover, the state of the specimen's bones suggested that it would have been six to eight years of age.

To definitely determine who (or what) the skeleton is, Nolan enlisted University of California, San Francisco pediatric genomics expert Atul Butte, MD, PhD.

Their full genetic analysis, which compared Ata's DNA sequences to those from both healthy and diseased references, proves that Ata is a female of South American descent and strongly implies that she was a preterm birth with a severe form of skeletal dysplasia and a bone-aging disorder that caused early growth-plate fusion (which normally doesn't occur until near adulthood).

"While the extraordinary phenotype of the specimen drove broad discussion as to its origin, and no hypothesis was left off the table during analysis, the specimen is shown here to have a purely earthly origin with mutations that reflect the visual determinations," the paper stated.

How a fetus harbored so many genetic defects, however, is likely to remain mysterious, considering that no one has yet come forward with insider knowledge.

The paper concluded, "While we can only speculate as to the cause for multiple mutations in Ata's genome, the specimen was found in La Noria, one of the Atacama Desert's many abandoned nitrate mining towns, which suggests a possible role for prenatal nitrate exposure, leading to DNA damage."

IFLSCIENCE

HOW LONG DOES IT TAKE TO POOP LEGO?

by Rosie McCall

HAVE YOU EVER WONDERED HOW LONG IT TAKES FOR a little yellow Lego person's head to pass through the human body? No, us neither. But an international team of pediatricians decided to find out anyway.

The answer: an average of 1.71 days.

In the name of science, six healthcare professionals volunteered to swallow a Lego head and spend the next few days sifting through their bowel movements to retrieve the evidence. To meet the grade for participation, the volunteers had never had gastrointestinal surgery and were able to demonstrate they could swallow such an object—and, perhaps most importantly of all, were fine about rummaging through their own poop.

The results have been published in the *Journal of Paediatrics and Child Health* in an article titled "Everything is awesome: Don't forget

the Lego." A reference, in case you haven't seen *The Lego Movie*, to the song "Everything Is AWESOME!!!" as well as the pediatricians' blog, *Don't Forget the Bubbles*.

To account for any individual differences, preingestion bowel habits were standardized by an appropriately named scale, the Stool Hardness and Transit (or

SHAT) score. The amount of time it took the head to travel from mouth to toilet was also aptly titled—the Found and Retrieved Time (aka the FART) score.

So, what did they find? It took an average of 1.71 days for the Lego head to exit the body, with a varied FART score between 1.14 and 3.04 days. The researchers also noted that "Females may be more accomplished at searching through their stools than males," adding that this "could not be statistically validated." Presumably, this is referring to the fact that one male volunteer never found his Lego head. Let's just hope it made it out OK.

Although extremely tongue in cheek, there is a point to this research. The team hope parents can now rest safe in the knowledge that their kids' extra-mealtime habits will likely not cause any nasty health complications.

"It is possible that childhood bowel transit time is fundamentally different from adult, but there is little evidence to support this, and if anything, it is likely that objects would pass faster in a more immature gut," the study authors wrote. "This will be of use to anxious parents who may worry that transit times may be prolonged and potentially painful for their children."

And if a Lego head or a similar object does go mysteriously AWOL, the pediatricians' advice to parents is not to go looking for it.

The study authors concluded, "If an experienced clinician with a PhD is unable to adequately find objects in his own stool, it seems clear that we should not be expecting parents to do so—the authors feel that national guidance could include this advice."

DOCTOR ISSUES WARNING OVER DANGEROUS AND DEADLY MASTURBATION—BUT DON'T WORRY, IT'S SAFE IF YOU DON'T DO THIS!

by Tom Hale

GONE ARE THE DAYS OF MASTURBATION BEING VIEWED as a shameful act, linked to madness, disease, and general nastiness. Nowadays, even the most prudish of medical professionals would agree that masturbation has numerous health benefits.

That said, there is harmless fun and then there are downright dangerous delights.

Speaking to the German daily newspaper *Bild*, forensic physician Dr. Herald Voss has warned that between 80 and 100 people in Germany die each year from risky masturbation practices.

The most common cause is oxygen deficiency, from autoerotic asphyxiation, where people strangle themselves during masturbation for an added sexual euphoria. The problem is that asphyxiation, being deprived of oxygen, can arise much faster and more easily than you think. Complete constriction of the carotid arteries, the major arteries that run up the side of your neck and supply oxygen to your brain, can leave you unconscious within 15 seconds and dead within minutes.

A separate study of autoerotic deaths in northern Germany found that the most common way to asphyxiate was through strangulation or hanging, although people often used plastic bags, masks, or drugs.

Most autoerotic deaths occur in men, according to Dr. Voss, although statistics are hazy, since many of the deaths are undocumented or misreported. In suspected autoerotic deaths, police typically look for clues, such as pornography, exposed genitalia, shackles, and lack of a suicide note. However, it's often easier to classify the death as a straightforward accident or a suicide.

"Relatives who find the body sometimes put things away, because the shame is so great," Dr. Voss told *Bild*.

A urology study from the US in 1985 set out to discover the prevalence of injuries caused by people masturbating with vacuum cleaners. It concluded: "Unfortunately, and contrary to apparent public appreciation, injury due to this form of autostimulation may not be unusual. Five cases of significant penile trauma resulting from this form of masturbation are presented, with a spectrum of severe injuries, including loss of the glans penis."

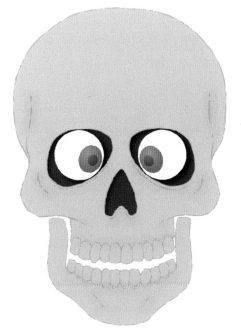

Voss added another story of a woman from the German city of Halle who reportedly found her son with Christmas tree lights clamped to his burnt body.

The moral of the story: Masturbation is a wonderful thing that everyone enjoys, but whatever your tastes or kinks, apply some common sense and stay safe.

A MAN TOOK WAAAAAAAY TOO MUCH VIAGRA. HERE'S WHAT HAPPENED TO HIM

by Rosie McCall

PRESCRIPTION DRUGS COME WITH RECOMMENDED DOSES for a very good reason, as one man has found out. The 31-year-old was admitted to an urgent care clinic with red-tinted vision two days after taking a little too much of the erectile-dysfunction medication Viagra. The condition, medics say, is irreversible.

In a first-of-its-kind study led by Mount Sinai, published in 2018 in the journal *Retinal Cases*, researchers have confirmed that high doses of sildenafil citrate (sold under the brand name Viagra) can damage your vision—and that the effects may be permanent. (Older research suggested the drug could cause permanent damage to vision in people with retinitis pigmentosa, but this was based on studies of mice.)

"People live by the philosophy that if a little bit is good, a lot is better," Richard Rosen, director of Retina Services at New York Eye and Ear Infirmary of Mount Sinai and lead investigator, said in a statement. "This study shows how dangerous a large dose of a commonly used medication can be."

Rosen and his team examined the retina of the 31-year-old man to check for structural damage right down to the cellular level (apparently, a world first). To do this, they used an electroretinogram, optimal coherence tomography, and adaptive optics, which lets scientists analyze microscopic optic structures in extremely high detail in real time. This meant they were able to pinpoint areas showing microscopic injuries to the cones in the retina, the very cells essential for color vision.

So, what did they find?

It was bad. The man's retina showed damage comparable to that found in animal models of hereditary retinal diseases like, for example, cone-rod dystrophy—which was unexpected, the researchers say.

"It explained the symptoms that the patient suffered from," Rosen added. "While we know colored vision disturbance is a well-described side effect of this medication, we have never been able to visualize the structural effect of the drug on the retina until now."

Before the experiment, the man admitted to taking much more than the recommended 50-milligram dose of a liquid sildenafil citrate he had bought online, telling medics the symptoms began to appear very shortly after. However, he wasn't able to specify exactly how much he had taken—instead of using the measuring pipette included in the pack, he drank the solution straight from the bottle.

While it is clearly a good idea to abide by medically approved guidelines, even standard doses of sildenafil citrate can cause "visual disturbances." Usually, however, the effect is to cast the world into a slightly bluish—not red—haze, and this should only be temporary, with symptoms usually subsiding within 24 hours, the researchers say.

As for the 31-year-old patient, at one year on since his first diagnosis, his vision had not improved. Treatments haven't helped and doctors conclude that the damage is permanent.

One Joint May Be All It Takes to Change the Structure of the Teenage Brain

by Rosie McCall

Smoking cannabis just once or twice may be enough to change the structure of brain areas linked to emotion and memory. So say researchers at the University of Vermont, who have been studying the effects of low-level drug use in a sample of 14-year-olds.

Apparently Penis Whitening Is a Thing

by Rosie McCall

The Lelux Hospital in Bangkok, Thailand, is offering penis whitening. The Thai health ministry has responded, however, warning that side effects include pain, inflammation, and scarring, and that patients who stop treatment will see their penis return to its usual shade, and could end up with some "nasty-looking spots."

WOMAN WHO RECEIVED LUNG TRANSPLANT DEVELOPED PEANUT ALLERGY FROM HER DONOR

by Katie Spalding

For one patient in 2018, a lifesaving lung transplant came with an unexpected price: a brand-new nut allergy.

Although the patient, a 53-year-old woman suffering from emphysema, didn't show the more common allergy symptoms, such as a skin rash or a stomachache, doctors realized that her breathing difficulties and chest tightness were the result of an allergy after she told them when it started—just after she chowed down on a delicious peanut butter and jelly sandwich.

The thing was, the woman had never had a peanut allergy before. But somebody else had—the 22-year-old man whose left lung had just been transplanted into her chest.

It's not the first time this has happened. In 2016, a 46-year-old man was given a kiwi fruit allergy along with a bone marrow donation from his sister. And in 2003, one liver transplant recipient inherited the very nut allergy that killed the donor.

Despite these examples, scientists aren't quite sure what causes the phenomenon.

IFLSCIENCE

WEED OR BOOZE? SCIENTISTS FINALLY SETTLE WHICH IS WORSE FOR YOUR BRAIN

by Madison Dapcevich

POTHEADS AND BOOZEHOUNDS HAVE BEEN DUKING it out for ages, but scientists have finally settled the age-old debate of which is worse for you—marijuana or alcohol.

And ... sorry, boozers. It turns out, marijuana may not be as damaging to the brain as previously thought.

Researchers examined the brains of more than 1,000 participants of varying ages by looking at neurological imaging data from MRI scans. Specifically, they used the data to examine the types of tissue that make up the brain: gray and white matter. Gray matter includes cell bodies that, among other things, enable functionality, while white matter allows everything to communicate. A loss of either would mean the brain wasn't working properly.

The team found that marijuana and cannabinoid products did not have long-term effects. Alcohol, on the other hand, was significantly associated with a decrease in gray matter size and white matter integrity, especially in adults with decades of exposure. The findings were published in the journal *Addiction* in 2018.

The negative impacts of alcohol on the brain have long been known, and it was assumed that cannabinoids were damaging to long-term brain health as well because of their immediate psychoactive effects.

"With alcohol, we've known it's bad for the brain for decades," said study co-author Kent Hutchison in a statement. "But for cannabis, we know so little."

A lot of past research studying the negative effects of marijuana came up with differing results, said Hutchison, who is a professor of behavioral neuroscience at the University of Colorado, Boulder.

"The point is that there's no consistency across all of these studies in terms of the actual brain structures," he said.

Researchers are quick to add that this doesn't mean pot is better for you or that the study proves any health benefit from toking up. It just means marijuana might be less harmful than previously believed.

"Particularly with marijuana use, there is still so much that we don't know about how it impacts the brain," said Rachel Thayer, a graduate student in clinical psychology at CU Boulder and the lead author of the study. "Research is still very limited in terms of whether marijuana use is harmful, or beneficial, to the brain."

The researchers say the study could help to better inform potential alternative pain treatments in the face of the ongoing opioid epidemic—the vast increase in the use of opioid-based analgesics.

Study Finds Spanked Children Are More Likely to Have Developmental Delays

by Robin Andrews

In January 2018, a study by the Donald and Barbara Zucker School of Medicine at Hofstra/Northwell found that children punished by spanking were five times more likely to experience developmental delays. The finding dovetails with other work suggesting that corporal punishment slows cognitive development and leads to antisocial behaviors.

Woman Develops Rare Condition That Leaves Her Unable to Hear Men

by Rosie McCall

A woman from Xiamen, in southeast China, woke up one morning to find that she couldn't hear a word her boyfriend, or any other men, were saying, reports the *Daily Mail*. She was later diagnosed with reverse-slope hearing loss (or RSHL), which leaves sufferers unable to hear low-frequency sounds.

HEALTH AND MEDICINE

THERE'S SOMETHING YOU NEED TO KNOW ABOUT THE MCDONALD'S TOUCHSCREENS
by Tom Hale

NEXT TIME YOU USE A SELF-ORDER TOUCHSCREEN AT McDonald's, you might also get some "free side orders" of Listeria, Staphylococcus, and a bunch of other bacteria found in poop. Mmmm, I'm lovin' it!

An investigation in 2018 found that McDonald's touchscreens are covered in a number of potentially worrying and contagious strains of bacteria.

Microbiologists from London Metropolitan University carried out a swab analysis of touchscreens in eight McDonald's restaurants in the UK—six in London and two in Birmingham—as part of an investigation by *Metro.co.uk*.

Samples from all of the branches contained coliform bacteria, a broad class of bacteria found in the poop of all warm-blooded animals and humans. They also discovered Listeria in two London branches. Typically associated with causing a food poisoning–like sickness, this can be potentially life-threatening if you're pregnant or have a weak immune system.

"Listeria is another rare bacterium we were shocked to find on touchscreen machines as, again, this can be very contagious," Dr. Paul Matewele, a London Metropolitan University microbiologist who worked on the investigation, told *Metro*.

A screen at one branch was found to harbor Staphylococcus, a group of bacteria that can be responsible for skin infections as well as food poisoning. Other bacteria discovered on the screens include Pseudomonas (responsible for chest infections), Enterococcus faecalis (found in the gastrointestinal tracts of humans), and Klebsiella (associated with urinary tract infections, septicemia, and diarrhea).

"Enterococcus faecalis is part of the flora of gastrointestinal tracts of healthy humans and other mammals. It is notorious in hospitals for causing hospital-acquired infections," added Dr. Matewele.

A McDonald's spokesman commented: "Our self-order screens are cleaned frequently throughout the day. All of our restaurants also provide facilities for customers to wash their hands before eating."

So just how worried should you be?

Well, bacteria are an inescapable part of our world. They are literally everywhere, from your face to the openings of deep-sea hydrothermal vents, or that ATM you got money out of this morning. Some of these are harmless, some are harmful, and some are actually beneficial.

Obviously, the idea of touching a slightly greasy screen covered in germs just before you eat is not very appetizing. However, as long as your immune system is in good shape, you shouldn't have too much to worry about. Many of these bacteria sound scarier than they are and, in small numbers, won't do you any harm.

That said, having a quick hand wash with soap and warm water before you eat anywhere in public probably isn't the worst idea in the world.

Bon appétit!

HEALTH AND MEDICINE

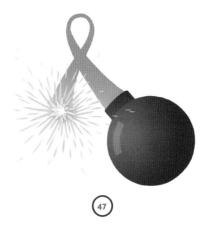

A CANCER "KILL SWITCH" HAS BEEN FOUND IN THE BODY—AND RESEARCHERS ARE ALREADY HARD AT WORK TO HARNESS IT

by Aliyah Kovner

AFTER EIGHT YEARS SPENT ANALYZING THE HUMAN genome and its many regulatory molecules, in 2018 a team from Northwestern University discovered a seemingly foolproof self-destruct pathway that can be used to knock out any type of cancer cell.

The mechanism involves the creation of small RNA molecules (called siRNAs) that interfere with multiple genes essential to the proliferation of fast-growing, malignant cells, but have little effect on normal, healthy cells. RNA, or ribonucleic acid, is normally used by the body to transmit genetic information from DNA to sites where new protein cells are created.

Research leader Marcus Peter and his colleagues characterized the fatal cascade of events these siRNA molecules trigger—dubbed DISE, for

"Death by Induced Survival Gene Elimination." They found DISE-associated genetic sequences were present in many naturally occurring RNA molecules in the body.

"We think this is how multicellular organisms eliminated cancer before the development of the adaptive immune system, which is about 500 million years old," Peter said in a statement. "It could be a fail-safe that forces rogue cells to commit suicide. We believe it is active in every cell protecting us from cancer."

Peter and his team observed the process by which our cells chop a larger RNA strand into multiple siRNAs. Remarkably, they found that about 3 percent of all our coding RNAs could be processed to serve this purpose.

"Now that we know the kill code, we can trigger the mechanism without having to use chemotherapy and without messing with the genome," Peter said in a press release.

The DISE pathway kills cancer cells in a brutal, simultaneous attack. "It's like committing suicide by stabbing yourself, shooting yourself, and jumping off a building all at the same time. You cannot survive," he explained in 2017. And all the research conducted thus far indicates that cancer cells cannot acquire resistance to DISE.

In a separate proof-of-concept study, published in *Oncotarget* in 2017, the Northwestern team used nanoparticles to deliver DISE siRNAs to the cells of human ovarian tumors that had been implanted in mice. The treatment resulted in a profound reduction in tumor growth without harmful side effects.

"Based on what we have learned in these [past several] studies, we can now design artificial microRNAs that are much more powerful in killing cancer cells than even the ones developed by nature," Peter concluded.

HEALTH AND MEDICINE

FIRST-EVER BABY BORN FOLLOWING A UTERUS TRANSPLANT FROM A DECEASED DONOR

by Tom Hale

IN 2018, THE FIRST-EVER BABY WAS BORN FOLLOWING a uterus transplant from a deceased donor. The case was documented in the medical journal *Lancet*. With live donors often in short supply, this breakthrough could offer hope to the 1 in 500 people who experience infertility problems arising from uterine anomalies.

The healthy baby girl was born via cesarean section at Hospital das Clínicas, at the University of São Paulo School of Medicine, Brazil, after her mother underwent a uterine transplant followed by an in vitro fertilization (IVF) pregnancy just seven months later.

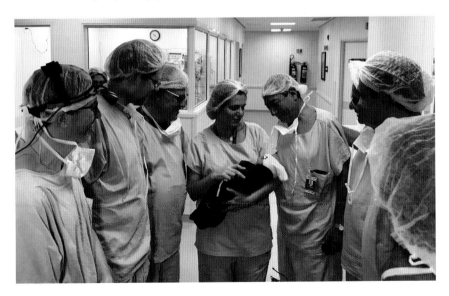

The mother, a 32-year-old woman, was born without a uterus, due to Mayer-Rokitansky-Küster-Hauser (MRKH) syndrome, a condition affecting 1 in 4,500 women. She received the new uterus during a 10 1/2-hour surgery in September 2016 from a 45-year-old donor who had died of a stroke. This uterus was later removed during the live birth to curb the chance of any issues with organ rejection.

The mother and child were able to leave the hospital just three days after the birth, and the following months were gloriously uneventful.

"The use of deceased donors could greatly broaden access to this treatment, and our results provide proof-of-concept for a new option for women with uterine infertility," said Dr. Dani Ejzenberg of the São Paulo School of Medicine, lead author of the report.

"We are authorized to do two more cases and we are focused on improving our protocol to be able to repeat this success story," Natalie Ehrmann Fusco, a spokesperson for the doctors on the project, told IFLScience.

There have been 11 births following uterine transplantation from living donors, the first of which took place in Sweden in September 2013. But all previous attempts to birth a child using a transplanted uterus from a deceased donor have proved unsuccessful.

As if this new case study is not impressive enough, it is also the first uterine transplantation of any kind to take place in Latin America.

The downside of the surgery is the high dose of immunosuppressant drugs, required to prevent the mother from rejecting the transplanted uterus, and the moderate, although manageable, levels of blood loss. The researchers state that transplants from deceased donors might have some benefits over donations from live donors. Most obviously, doctors do not have to worry about the donor's health. They also note that it could potentially make uterus transplants more widely available, especially in countries that already regulate and distribute transplant organs from deceased donors.

"The first uterus transplants from live donors were a medical milestone," added Dr. Ejzenberg. "However, the need for a live donor is a major limitation, as donors are rare, typically being willing and eligible family members or close friends. The numbers of people willing and committed to donate organs upon their own deaths are far larger than those of live donors, offering a much wider potential donor population."

Hand Dryers Spread Bacteria So Dramatically That Scientists Think They're a Public Health Threat

by Aliyah Kovner

Researchers from the University of Leeds, UK, have found that the no-touch jet-air dryers in public restrooms blast bacteria from people's poorly washed hands into the air and onto nearby surfaces in disturbing quantities. They introduce 27 times more bacteria into the air, which then circulate for 15 minutes afterwards.

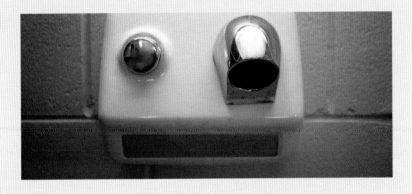

CHAPTER FOUR

PLANTS AND ANIMALS

NOT ALL OF US ARE BLESSED WITH DAVID ATTENBOROUGH'S khaki pants and keen botanical acumen. Luckily, our Plants & Animals chapter is here to keep you informed about the latest, most important, and downright piss-yourself-hilarious developments from the scientific study of the natural world.

No one's quite sure how life got started on Earth, though there are plenty of theories. The "RNA world" hypothesis, for example, says that life spontaneously emerged from RNA—a molecule a bit like DNA—which is capable of both storing the genes that encode life and making copies of itself. There are other ideas. Some scientists even believe in a concept known as "panspermia," which says that life on Earth may have actually begun elsewhere in the universe and was then transported to our planet inside comets and asteroids. There is certainly evidence that microbes can survive the harsh vacuum of space for extended periods.

What scientists do more or less agree on is how those primitive early organisms grew and developed into the diverse plethora of complex living things that our planet is teeming with today. In 1859, Charles Darwin published his theory of evolution by natural selection. Its core premise is that species adapt to their environment. Random mutations creep in from one generation to the next, some of which make a species better able to thrive in its surroundings—and these ultimate-survivor organisms are then more likely to pass the beneficial traits down to their offspring. Rinse and repeat.

But the environment is the key. Put some bugs in the desert and they'll evolve into new forms of life that are suited to the arid conditions. Chuck the same bugs in the ocean and, after many generations, very different life-forms will emerge.

One controversial consequence of this is that modern humans are now thought to have descended from primates, our closest living relatives being chimpanzees and their cousins, bonobos. We became a species in our own right, *Homo sapiens*, in Africa several hundred thousand years ago, before migrating to the rest of the world.

I say "controversial," though that only applies if you believe that human beings miraculously popped up one day in the Garden of Eden. "Primate change deniers," or creationists, as they're more generally known, believe that the Earth is less than 10,000 years old and that life here was fashioned in seven days by a bearded man on a cloud—despite the wealth of evidence to the contrary in the fossil record. (Well, just to be clear, there's actually nothing in the fossil record to indicate that God, if he does exist, isn't a bearded man—though there's also nothing to suggest that he exists in the first place.)

Evolution is still driving the development of species today. Although, in the case of human beings, we're altering our environment and our bodies at a much faster pace than natural adaptation can keep up with. One consequence of this is that some traits we've evolved are now woefully mismatched to the environment we've created for ourselves. For example, our natural cravings for calorie-rich, sugary, and fatty foods, which evolved at a time when these commodities were scarce, now do us very few favors in the modern age of plenty. Indeed, they have created an obesity epidemic. Futurists have suggested that we could turn the situation on its head, using technologies such as genetic modification and nanotech to take control and steer our own evolutionary path, tailoring our species to become "transhuman."

For now, though, we'll have to make do with mother nature, which is both beautiful and cruel. Coming up, we report on the shocking "Chimp civil war" that broke out between two rival factions of chimpanzees living in Gombe National Park, Tanzania. Observed and documented by naturalist Jane Goodall, the war was brutally violent, and didn't stop until one side had completely eradicated the other. More proof, as if it were needed, that humans and great apes really are cut from the same cloth.

You can also find out about the smallest dinosaur footprints, dating from about 110 million years ago and measuring just half an inch (1 cm) across, found in South Korea. And check out the discovery that spiders actually nurse their young with milk (a trait previously thought to be almost exclusive to mammals).

Of course, there's also the usual grab bag of "human interest" stories. Wondering why your hangovers are getting worse? It could just be that evolution is cranking up the symptoms to warn us off the booze and thus guard us from its long-term health effects. Survival of the fittest and all that. Though perhaps it's for the best—other research is suggesting that the majority of coffee plants could be heading for extinction. Catch up on both stories here.

Have you ever seen wombat doo-doo? If you have, you'll know that it's rather unusual for the simple reason that it's cubic in shape. Apparently, this allows the wombat to stack its poo up and thus mark its territory on the slopes where it lives—without the poo rolling away. Clever, eh? Recently, scientists embarked on a landmark study of wombat bowels to figure out exactly how they achieve this remarkable feat.

So, from now on, should you ever happen upon a wombat pulling a strained face reminiscent of a crooning boy-band singer trying to shit a small car . . . you'll know why.

PLANTS AND ANIMALS

BRUTAL CHIMPANZEE WAR WAS LIKELY DRIVEN BY POWER, AMBITION, AND JEALOUSY

by Josh Davis

WHEN A NEW LEADER WAS CROWNED, IT WAS HOPED that the community would settle down and peace would prevail. But two younger pretenders had other ideas, their lofty ambitions luring them to try to seize power for themselves. The resulting fracture in the group led to years of brutal warfare, during which raids were conducted, ambushes set, and no one was above murder.

No, that's not a plot outline from *Game of Thrones*, but an account of the only known fully documented chimpanzee civil war. The conflict became known as the Four-Year War of Gombe—proof, if it was needed, that chimps and humans are two of a kind. A study has reexamined the episodes that led up to the war, to figure out what sparked it.

The events were recorded by English primatologist Jane Goodall after a decade of watching the community of chimps at Gombe National Park, in Tanzania, at a time when chimpanzees were still thought to be peaceful, forest-living apes. Between the years of 1974 and 1978, she observed the extreme violence that erupted as the one community seemingly split and the apes waged a savage war. What she witnessed truly disturbed her.

"Often when I woke in the night, horrific pictures sprang unbidden to my mind—Satan [one of the apes], cupping his hand below Sniff's chin to drink the blood that welled from a great wound on his face.... Jomeo tearing a strip of skin from Dé's thigh; Figan, charging and hitting, again and again, the stricken, quivering body of Goliath, one of his childhood heroes," Goodall wrote in her memoir.

But the cause of the war has always been up for debate. Was it a natural event that was occurring independently of Goodall, who was simply observing the apes, or was it sparked by the feeding station that she had set up in the forest?

After digitizing all of Goodall's original field notes from her time at Gombe and then sifting through them, a US team of researchers built up an impressively detailed picture of the social interactions and friendships between the chimps, and mapped how these changed. They published their results in 2018 in the *American Journal of Physical Anthropology*.

They found that the seeds for the conflict had been present for some years. While at the end of the 1960s, all males were intermingling quite happily, by 1971, fractures in the community were beginning to emerge. The northern and southern males were starting to spend less time with each other, and encounters were becoming increasingly aggressive.

Within a year, the two sides had become distinct, with the chimpanzees staying and socializing only within their own groups, a full two years before the fighting spilled over into all-out warfare. The researchers suspect that the divide occurred after an ape called Humphrey became the alpha male, something the southern males Charlie and Hugh disagreed with.

Over the next four years, and a campaign of skirmishes, violence, and kidnapping, Humphrey and his northern community killed every single male in the southern group and took over their territory, as well as the only three surviving females. In fact, this latest study shows that it was likely the limited number of mature females in the forest at the time that precipitated the conflict.

The researchers suggest that—not unlike the behavior we see in some human communities today—the infighting among the males was largely driven by ambition, jealousy, and the desire for power. As such, it would likely have occurred with or without Goodall being there.

EVOLUTION COULD DESTROY OUR ABILITY TO TOLERATE ALCOHOL
by Rosie McCall

RECENT STUDIES HAVE SHOWN THAT WE ARE STILL evolving. Natural selection picks out genes that make each new generation of humans stronger and fitter than the last, and better equipped to survive in its environment.

Research published in *Nature Ecology and Evolution* suggests we could even be evolving a gene that diminishes our tolerance for alcohol.

Researchers at the University of Pennsylvania analyzed the genomes of some 2,500 people from 26 populations across four continents, using data collected by the 1000 Genomes Project, an international collaboration to study human genetic diversity. The team then looked for specific genetic traits that appear in disparate populations. They might find, for example, that a particular set of mutations has occurred in populations located in parts of both Africa and Asia. For this to have happened, the mutation must have emerged independently in these two different populations and persisted.

Interestingly, one such emerging group of mutations were modifications to the genes controlling alcohol dehydrogenase (ADH), which could change how our bodies process alcohol. ADH is the enzyme that breaks down alcohol, which it does by metabolizing it into a compound called acetaldehyde. This toxic chemical is responsible for your hangover the morning after a night out. Fortunately, the body is able to turn this nasty stuff into the non-toxic substance acetate relatively quickly, and we feel better after a day or so.

However, evolution could be finding a way to curb humanity's love of alcohol by creating new variants of ADH that affect our tolerance for booze and our body's ability to process acetaldehyde. Essentially, it means we would feel ill after just a small amount of drink.

So far, these genes have only been detected in East Asia and West Africa, but time will tell how far they spread.

The National Institute on Alcohol Abuse and Alcoholism estimates that, each year in the US, 88,000 people die from alcohol-related causes. By imbuing humans with an aversion to alcohol, evolution may actually be boosting our survival chances in an altogether unexpected way.

DON'T THINK ARACHNIDS ARE LOVING? THIS SPIDER NURSES ITS YOUNG WITH MILKY LIQUID

by Madison Dapcevich

DESPITE THEIR CREEPY, CRAWLY APPEARANCE, SOME species of spiders make for exceptionally caring parents. Even more impressive, Toxeus magnus, a species of jumping spider, nurses its young much as mammals do, feeding its spiderlings nutritious milk packed with four times the amount of protein found in cow's milk, according to a study published in November 2018 in the journal *Science*.

Surprised? You're not alone. Researchers at the Chinese Academy of Sciences first became interested in the odd behavior of *Toxeus magnus* when they noticed that the spiders nest in the same way ants do, creating a space to shelter several spiders at a time.

"It's a puzzling observation for a species assumed to be non-colonial," said study author Chen Zhanqi, of the Chinese Academy of Sciences, in a statement. "It's possible that the jumping spider might provide either prolonged maternal care or delayed dispersal. We decided to test it."

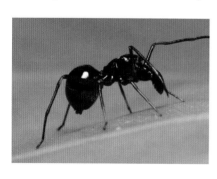

The mother spider wasn't seen to bring food back to the nest, but her babies continued to grow. So, the team did what any responsible scientist would do and grabbed a few microscopes. Upon closer observation, they could see droplets of nutritious fluid "leaking from the mother's epigastric furrow"—a specially designed sexual organ found on the abdomen. Mom would deposit these milk droplets on the nest, where her babies could then come and suck them up. After the first week, she stopped depositing the droplets and instead allowed the spiderlings to suck directly from her. When researchers blocked milk production, the spiders stopped developing and died, showing their "complete dependence on the milk supply."

This behavior was observed both in a laboratory and a field setting, and lasted until the spiders were at least 20 days old (at which point they are able to feed themselves) and generally until the spiders had reached their sub-adult stage at around 40 days. During this time, Mom was also seen taking care of the nest and helping her babies to shed their exoskeletons.

When given both maternal care and milk, more than three-quarters of the hatched offspring survived to adulthood and reached a normal body size. Though Mom treated all her babes the same, she only allowed her daughters to return to the nest after reaching sexual maturity. Adult sons were attacked if they tried to come home, probably to reduce the likelihood of inbreeding.

Lactation is a defining characteristic of mammals—and there are only a few known cases of it in other classes of the animal kingdom. The researchers hypothesize that arachnids most likely evolved the ability to lactate in response to predation risk, uncertain food access, and as a way to survive in harsh living environments.

"Our findings demonstrate that mammal-like milk provisioning and parental care for sexually mature offspring have also evolved in invertebrates," said Chen. "We anticipate that our findings will encourage a reevaluation of the evolution of lactation and extended parental care and their occurrences across the animal kingdom."

A SCIENTIST HAS BEEN EATEN ALIVE BY A CROCODILE

by Tom Hale

A SCIENTIST WAS EATEN ALIVE BY A CROCODILE IN JANUARY 2019 after falling into an enclosure at a research facility in Indonesia.

The body of Deasy Tuwo, a 44-year-old woman, was found on the morning of January 11, 2019 in an outside pool at CV Yosiki Laboratory in North Sulawesi, Indonesia. Police say a 17-foot-long (5-meter-long) crocodile leapt up against the wall of the enclosure during feeding time and grabbed the researcher, pulling her into the pool and eating parts of her body.

"We were curious when we looked at the crocodile pool. There was a floating object; it was Deasy's body," said Erling Rumengan, a colleague of Deasy, according to the UK newspaper the *Mirror*.

"She was the head of the lab. A quiet person. We're confused about how this has happened," said another colleague.

The fate of the crocodile is unknown, however, photographs from the scene show the animal tied to a truck by a huge crowd of people. Local media reports claim the croc was being transported to a wildlife center in Bitung district for tests to be carried out on its stomach contents. Only half of the scientist's remains were found in its enclosure.

Crocodile attacks are often fatal. Worldwide, crocodiles are estimated to kill about 1,000 humans per year. Most of these deaths are caused by the Nile crocodile and the saltwater crocodile, largely because these species live near human populations.

GM Crops Found to Increase Yields and Reduce Harmful Toxins in 21 Years of Data

by Jonathan O'Callaghan

A study of genetically modified crops has found that they increase food production and can also be good for you. Researchers from the Institute of Life Sciences, in Italy, conducted a meta-analysis of 6,006 peer-reviewed studies from 1996 to 2016 on GM maize, finding greater yields and lower concentrations of toxins.

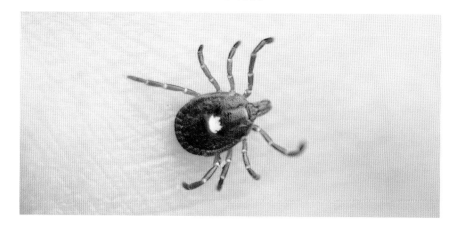

THE "REVERSE ZOMBIE" TICK IS SPREADING AROUND AMERICA, CAUSING A STRANGE CONDITION AS IT GOES

by James Felton

A RARE BREED OF TICK THAT CAN CAUSE EVERYTHING from intense itchiness and stomach cramping to difficulty breathing and even death has been spreading across the USA. The tick is also spreading something far stranger as it goes: an allergy to meat.

The lone star tick, aka the "reverse zombie" tick, makes you shy away from meat, rather than craving it (as a bite from a zombie might). One bite from the tick, in fact, and you can develop a life-threatening allergy to a sugar molecule found in red meat.

Once you've been bitten, your immune system can become triggered by the presence of galactose-alpha-1,3-galactose (alpha gal) and go into overdrive—meaning the next time you eat meat from a mammal that produces this sugar (e.g., pork and beef), you may find yourself breaking out in massive hives or going into anaphylactic shock.

The condition was only recently discovered in 2004 when immunologist Thomas Platts-Mills found that a group of cancer patients were all suffering from the same symptoms, *Wired* reports.

The patients were all on the same drug—Cetuximab—but those living in the southeastern USA were 10 times more likely to report symptoms such as itching, swelling, and low blood pressure. This was quite strange, as you wouldn't expect symptoms to be so area-specific.

Platts-Mills began to investigate the blood of the symptomatic patients, and found antibodies for alpha gal—the sugar found in meat. Cetuximab is full of the sugar, as it is derived from genetically modified mice. Eventually, he discovered that 80 percent of the allergic patients also reported having been bitten by a tick. Platts-Mills has since shown that bites from the lone star tick lead to a 20-fold increase in alpha gal antibodies. Researchers are now trying to figure out why saliva from the ticks causes the immune system to attack alpha gal as a foreign body.

"There's a lot we don't know about the allergy," Dave Neitzel, an epidemiologist with the Minnesota Department of Health, told the *Herald Review*. Some people who get bitten by the ticks don't develop the allergy at all.

"There's something really special about this tick," Jeff Wilson, an asthma, allergy, and immunology fellow in Platts-Mills's group, told *Wired*. "Just a few bites and you can render anyone really, really allergic."

While the team investigates why the tick causes a meat allergy, the only way to protect yourself is to use bug repellent in areas where the lone star tick dwells.

IFLSCIENCE

THIS IS WHAT EATING PEOPLE DOES TO THE HUMAN BODY

by Tom Hale

CANNIBALISM IS ONE OF THE DARKEST TABOOS IN MANY cultures. But beyond the social stigma of eating fellow humans, the practice brings with it a strange, and fascinating, danger.

In 1961, a young Australian medical researcher called Michael Alpers headed to the Eastern Highlands of Papua New Guinea, inspired to merge his two passions of medicine and adventure. Here, he began to investigate a mysterious condition suffered by the Fore people, a scarcely touched tribe that lived deep in the mountains and practiced cannibalism.

"The body was eaten out of love as well as for gastronomic appreciation," Alpers wrote in one of his academic texts about the Fore people.

They called the condition "kuru." Every year, kuru would kill up to 200 members of the tribe, sometimes in startling circumstances.

Starting with tremors and an impaired ability to work, sufferers went on to develop a total loss of bodily function, depression, and often emotional instability, which sometimes exhibited itself as hysterical laughter. When word of the disease spread to the West, the media sensationally dubbed it the "laughing death."

The Fore people believed it was a terrible curse, but Alpers wanted to find a more scientific explanation. Curiously, the condition did not appear to be caused by a virus, bacteria, fungus, or parasite. Equally strange, it was only women and children who fell sick.

This made the researchers begin to wonder: Perhaps it had something to do with the Fore's funerary ritual of cannibalism. The practice involved only the women and children eating the brains, while the men would just eat the flesh.

By 1966, Alpers and a team of other scientists were starting to catch on to the fact that kuru was caused by so-called "prions." This discovery paved the way for Baruch S. Blumberg and D. Carleton Gajdusek to sweep up a Nobel Prize in Physiology or Medicine in 1976 for "their discoveries concerning new mechanisms for the origin and dissemination of infectious diseases."

Prions are proteins that have become twisted and turned to the "dark side." These infectious agents lose their functions and acquire the ability to turn other normal proteins into prions, too, thereby becoming infectious.

Some of the more infamous diseases caused by prions are BSE, aka "mad cow disease," and its human equivalent variant Creutzfeldt-Jakob disease (vCJD)—two degenerative brain disorders that share an uncanny resemblance to kuru. It's believed that BSE is most likely the result of cows eating nervous system tissue of other cows that has been recycled into feed.

So, eating human brains might not always be the best of ideas, even before you get into the whole array of blood-borne illnesses that you could contract. However, here's where the story takes a turn. A study published in *Nature* in 2015 found that the Fore people who regularly ate brains had developed a resistance to prion diseases, a discovery that is now helping scientists understand degenerative brain diseases, such as mad cow disease, vCJD, and some cases of dementia.

What doesn't kill you, it seems, really does make you stronger.

Mountain Gorillas Are No Longer "Critically Endangered" after a Successful Conservation Effort

by Jonathan O'Callaghan

The International Union for Conservation of Nature (IUCN) announced in 2018 that they were updating the status of mountain gorillas from "Critically Endangered" to simply "Endangered," after a campaign to bring them back from the edge of extinction had increased the number of these animals in the wild to more than 1,000.

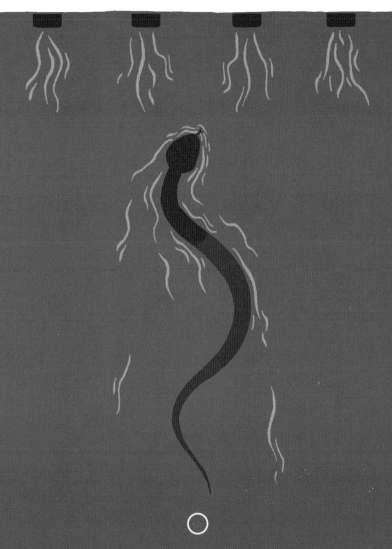

Zoo Creates World's First Reptile Swim-Gym to Fight Snake Obesity

by Stephen Luntz

No one likes a flabby snake. So Melbourne Zoo has found a way to keep its reptiles buff with the creation of a water gym that zoo officials think is the first in the world. The basic setup is a watery version of a treadmill, with the flow rate adjustable to suit the pace of the animal.

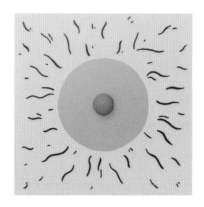

WHY DO MEN HAVE NIPPLES?
by Rosie McCall

TO PUT IT BLUNTLY, CERTAIN THINGS IN LIFE ARE UTTERLY pointless—pet rocks, for instance. See also: shoe umbrellas, goldfish walkers, diet water, and . . . male nipples.

In fact, the very existence of male nipples is such a conundrum that "Why do men have nipples?" racked up an average of 22,000 monthly Google searches in 2016. It appears that we are just as clueless as Erasmus Darwin (grandfather to Charles), who pondered the very same question back in the 18th century.

However, while there might not be a purpose as such, there is a biological reason why men have nipples. And it all comes down to embryonic development in the womb.

In the first few weeks after conception, the male and female embryo follow the exact same developmental path. It is not until the sixth to seventh week of gestation that reproductive organs start to develop and the fetus begins to differentiate by sex.

Specifically, this is when a gene called SRY starts to take effect. The SRY gene is basically an instruction manual for the production of the "sex-determining region Y protein," which initiates the development of the testes in male fetuses. Once these have formed—at around the nine-week mark—the male body starts to generate testosterone and the fetus begins to display more and more sex-specific biological characteristics.

But by this point, it is too late. At least as far as nipples are concerned, because the mammary glands begin to develop in the very first few weeks of gestation and before the SRY gene has had a chance to kick in. This means that while male nipples may be smaller than female nipples, they do still exist.

Admittedly, this doesn't explain why men have retained such a seemingly useless body part—if it's biologically useless, why haven't they evolved to go without?

It could be that male nipples do serve a purpose, just one that we don't quite understand yet. That purpose could be sexual (like women, many men can enjoy nipple stimulation) or social (though extremely uncommon in most societies today, men in the Aka Pygmy tribe in central Africa are known to nurse their babies). Or, perhaps, even biological. There are cases of men lactating, though this is usually in response to starvation or hormone imbalances and is very, very rare. That is to say, if it happens to you, get it checked out ASAP.

The most likely explanation, scientists say, is that male nipples aren't detrimental in any way, so removing them just isn't a priority, evolutionarily speaking. Like wisdom teeth, the appendix, and Darwin's point (the vestigial remnant of the pointed primate ear that some people have), the male nipple is just another quirk of evolution that we're stuck with.

IFLSCIENCE

WHALES BECAME REALLY STRESSED DURING WORLD WAR II, STUDY SHOWS

by Jonathan O'Callaghan

RESEARCHERS HAVE USED A RATHER INTRIGUING METHOD to find out what makes whales stressed, and—surprise, surprise—it looks like we're to blame.

In a study published in the journal *Nature Communications* in November 2018, a team led by Baylor University in Texas analyzed the earwax of fin, humpback, and blue whales living in the Pacific and Atlantic Oceans between 1870 and 2016. These species are all "baleen whales," distinguished by the double blowhole on the tops of their heads, and this is the first study to look at stress over time in these creatures.

Earwax accumulates in whales, forming a plug of material that preserves a chemical record of each animal's life. Analyzing it layer by layer, the researchers were able to measure the concentration of cortisol—which is a stress-response hormone—and determine how it had changed with time. And this enabled them to match up the whales' stress levels with key moments in history.

Amazingly, the researchers were able to show that cortisol levels were at their highest in the 1960s, when whaling was at its peak—and up to 150,000 of the animals were "harvested." The animals were also highly stressed during periods of increased whaling in the 1920s and 1930s.

But cortisol levels were also found to increase during World War II. Despite whaling activities actually declining during the conflict, the

researchers believe that the war itself could have caused the whales to become stressed.

"The stressors associated with activities specific to World War II may supplant the stressors associated with industrial whaling for baleen whales," Dr. Sascha Usenko, one of the study's co-authors, said in a statement.

"We surmised that wartime activities, such as underwater detonation, naval battles including ships, planes, and submarines, as well as increased vessel numbers, contributed to increased cortisol concentrations during this period of reduced whaling."

Whale cortisol levels reached their lowest point in the mid-1970s, when whaling decreased to reportedly zero in the Northern Hemisphere. However, they have increased steadily from then to the present day, suggesting that other stresses, perhaps including climate change, may be playing a part.

"While the generated stress profile spans nearly 150 years, we show that these whales experienced survivor stress," said Dr. Stephen Trumble, the study's lead author, in a statement. "Exposure to the indirect effects of whaling, including ship noise, ship proximity, and constant harassment, results in elevated stress hormones in whales."

Majority of Coffee Species Threatened with Extinction

by Rachel Baxter

According to a 2019 study from the UK's Royal Botanic Gardens (RBC), Kew, 60 percent of wild coffee species are threatened with extinction, as a result of climate change. Although we get our coffee from cultivated crops, we rely on wild coffee plants to sustain them through cross-pollination.

Cats Are Not Inherently Antisocial Creatures. It's Just You

by Rosie McCall

Researchers at Oregon State University recruited 46 cats to see how well they coped with human company. Perhaps unsurprisingly, the animals preferred spending time with people who were enthusiastic and attentive, as opposed to those who ignored them. The research was published in the journal *Behavioral Processes* in 2019.

WORLD'S SMALLEST DINOSAUR FOOTPRINTS FOUND, MEASURING LESS THAN HALF AN INCH

by Katy Evans

THE WORLD'S SMALLEST DINOSAUR FOOTPRINTS HAVE been found in South Korea, measuring a tiny 0.4 inch (1 cm) long, making this unknown species of raptor the size of a sparrow.

The tiny footprints date back to around 110 million years ago, when dinosaurs shared the Earth with both mammals and birds. The tracks were found in a dried lake bed in Jinju City, South Korea, which has yielded an abundance of other Cretaceous-period creatures, from birds and pterosaurs to crocodilians and mammals.

"These new tracks are just 0.4 inch (1 cm) in length, which means the dinosaur that made them was an animal you could have easily held in your hand," said Dr. Anthony Romilio of the University of Queensland in a statement. "They are the world's smallest dinosaur tracks."

With their characteristic three-clawed shape, the new footprints were immediately identified as those of a raptor—a bird-like, carnivorous, theropod dinosaur, made famous by its oversize depiction in the *Jurassic Park* movies. The researchers, however, are unsure of the particular species, or even if they were adults or juveniles.

"We do have tiny raptors known from fossil bones from China. Fossil bones of diminutive adult raptors called Microraptor were about the size of crows, with feet about 1 inch (2.5 cm) long," Dr. Romilio told IFLScience. "Even though Microraptor was very small (by dinosaur standards) it was still too large for our 0.4-inch (1-cm) tiny South Korean tracks. So perhaps this favors the tracks being made by baby raptors."

The species has been assigned to a new "dromaeosaurid ichnogenus," dubbed Dromaeosauriformipes rarus, which means "rare footprints made by a member of the raptor family known as dromaeosaurs." Dromaeosaurids are a family of small- to medium-sized, feathered therapods. And ichnogenera, which means "footprint group," are any genus that is only known through trace fossils, such as fossilized footprints, rather than actual remains.

Dr. Romilio has created a reconstruction of what Dromaeosauriformipes may have looked like. "I have covered them in downy feathers, with bold horizontal striping to be highly visible to each other, and maybe to be easily recognizable by a possible parent raptor," he explained, adding that this is very much a best guess, and by no means ironclad. The work was published in the journal *Scientific Reports*.

(64)

VERY GOOD PUPPY DIGS UP 13,000-YEAR-OLD MAMMOTH FOSSIL IN ITS OWNER'S BACKYARD

by Aliyah Kovner.

In September 2018, a Labrador retriever named Scout dug up a 13,000-year-old fossilized woolly mammoth tooth.

As reported by the Seattle-based outlet *Komo News*, owner Kirk Lacewell noticed that Scout was carrying something around in his mouth after he dug a shallow hole in the fenced backyard, on Whidbey Island, Washington.

After washing and drying Scout's find, Lacewell realized that it looked a lot like a bone. He sent pictures to paleontologists at the University of Washington's Burke Museum, where the scientists quickly agreed that the object was a mammoth tooth and estimated its age at around 13,000 years.

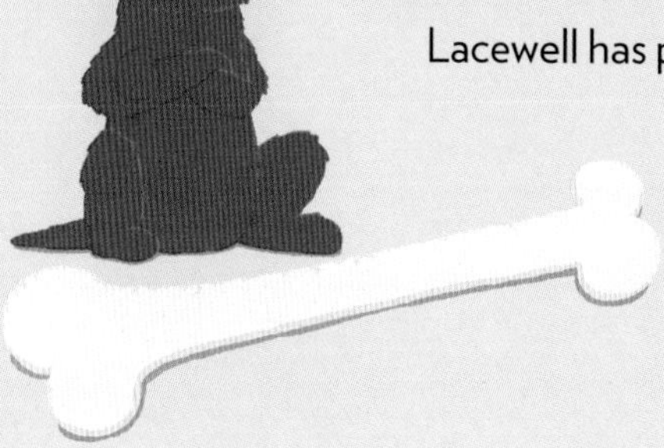

Whidbey Island was home to a large population of woolly mammoths before the species went extinct at the end of the last ice age, approximately 11,000 years ago.

Lacewell has put the tooth on his living room mantle, a place of honor that is conveniently out of the dog's reach.

PLANTS AND ANIMALS

WE NOW KNOW HOW WOMBATS PRODUCE THEIR UNIQUE CUBIC POOS

by Stephen Luntz

LIKE OTHER HERBIVORES, WOMBATS POOP A LOT, BUT unlike any other known species, their droppings are almost cubic, the size and shape of dice. Biologists have long had an explanation for *why* they do this, but most recently they have also figured out *how* the wombat digestive tract achieves this defecatory feat.

Any trait exhibited by one animal species alone among all the millions on Earth is interesting. The shape of wombat droppings is thought to help the animals mark their territory, by allowing them to produce tall piles that don't roll down the often steep hillsides of their habitat.

However, many other species use dung to mark their territory, and have never come up with this useful adaptation, for the simple reason that it isn't easy to construct a digestive system that produces cubic, rather than broadly cylindrical shapes. However, the wombat's secret has finally been cracked and presented at the American Physical Society's Division of Fluid Dynamics 2018 conference, held in Atlanta, Georgia.

The work was led by Dr. Patricia Yang of the Georgia Institute of Technology. "The first thing that drove me to this is that I have never seen anything this weird in biology," Yang said in a state-

[125]

ment. "I didn't even believe it was true at the beginning. I Googled it and saw a lot about cube-shaped wombat poop, but I was skeptical."

So Yang, who studies fluid dynamics within the body, did the appropriately scientific thing. She obtained wombats' digestive systems (from animals killed by cars) and inflated their intestines. The widespread assumption that wombat anuses must be square has long been debunked, and Yang also contradicted the previous theory that the cubic shape is formed at the top of the intestine.

Instead, the contents of the wombat's stomach come down the gut in a semi-liquid state, only to solidify in the last 8 percent of the intestine. Here, alternating rigid and flexible stretches of the intestine walls apply very different strains to the incipient stools, making corners and edges, and producing the characteristic cubic shapes.

Curiosity-inspired research always attracts charges of wasting taxpayer funds, but Yang thinks there could be a practical payoff. "We currently have only two methods to manufacture cubes: we mold it, or we cut it. Now we have this third method," she said. Whether replica wombat intestines will prove advantageous in manufacturing, however, remains to be seen.

One thing Yang hasn't investigated yet is the question of just how painful the whole process is for wombats, who are constantly shitting out small bricks.

Step Aside Knickers, There's an Even Bigger Cow in Town Called Dozer

by Jonathan O'Callaghan

Knickers, Australia's giant 6.4-foot-tall (1.95-meter) cow that made news headlines in 2018 has now been beaten by an even bigger cow from Canada. Called Dozer, this bovine measures in at 6.5 feet (1.98 meters). However, both of them fall short of the record, held by an Italian ox called Bellino, which measures 6.8 feet (2.07 meters).

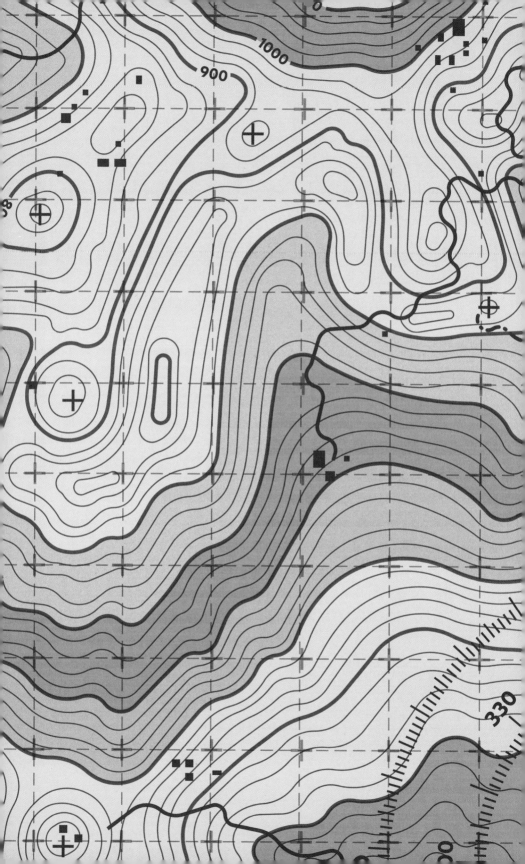

CHAPTER FIVE
ENVIRONMENT

UNLESS YOU'VE HAD YOUR HEAD UNDER A ROCK since the end of the last ice age, you'll probably be aware that things are not altogether well with our planet.

To cut a very long story quite short, global temperatures have, since the end of the industrial revolution in the mid-19th century, risen by about 0.8°C/1.4°F. That might not sound like very much, but the Intergovernmental Panel on Climate Change (IPCC) predicts a further temperature rise this century of around 4°C/7°F, melting much of the Earth's polar ice and raising average sea levels by about 1 1/2 feet (46 cm). That's based on the current trend (the reality could be much worse—or much better, if we act). As it stands, it's sufficient to redraw the world's coastlines and displace millions of people from their homes in low-lying coastal locations.

All the evidence points to the cause being carbon dioxide released into the atmosphere by human industry. Carbon dioxide, and other so-called "greenhouse gases," cause global warming by trapping infrared heat radiation, preventing the heat from being released to cool the planet down. If radiation arrived from the Sun only at infrared wavelengths, there would be no warming—in fact, we'd have the opposite problem, as the Sun would be unable to provide enough heat and the planet would steadily cool and gradually slip into an ice age. But that's not what happens. Instead, radiation arrives at all wavelengths, many of which pass straight through the atmosphere unimpeded, warming the ground below. The trouble is the heat is then re-emitted by the ground as infrared, and this then gets trapped. The result is global warming.

When the first detailed calculations of global warming were produced, by the Swedish chemist Svante Arrhenius in 1896, it was seen as a good thing—a way to stave off the next ice age. However, over the course of the 20th century, evidence gradually accrued to support the conclusion that the long-term effects would be anything but beneficial. By the late 1980s, a significant number of scientists were calling for reductions in human-made pollution in order to avert what they saw as a major threat to the security of the planet.

Most human carbon dioxide emissions—at least, the problematic ones—are caused by the burning of fossil fuels. These are combustible chemicals formed from the remains of prehistoric plants and animals mined from deep underground. They're rich in carbon, which is released as CO_2 when the fuels are burned.

Living things absorb carbon. Plants, for example, soak up carbon dioxide from the atmosphere, through the process of photosynthesis, locking the carbon into their structure and releasing oxygen—which is handy for us and other animal life to breathe. When these organisms die and are buried underground, their carbon content gets buried with them. But while the burning of wood and other biofuels simply recycles existing carbon from organisms to the atmosphere and back again, the use of fossil fuels is dredging up old carbon, which was safely buried underground, and dumping it back into the atmosphere. This increase in CO_2 is trapping more heat, and that's why the planet is warming up.

Of course, global warming is an average trend. There will still be periods that are colder than average—this is the careful distinction we draw between day-to-day variations, which we call "weather," and the long-term climate. And that's why it's very much possible for, say, temperatures during winter in Chicago to hit −30°F, even though the global temperature trend is going in the opposite direction.

Some of the latest developments in the understanding of our changing climate are presented in the following pages. At the end of 2018, scientists confirmed that the previous four years had been the warmest on record. And this has led the rate at which Antarctic ice is melting to increase sixfold from 1979 to the present. Antarctica is now shedding over

250 billion tons (227 billion tonnes) of ice every year, all of which is driving sea levels higher.

In other environmental news, we report on the giant fatberg that was found in the sewers beneath the British seaside town of Sidmouth. You can find out why Earth's magnetic field appears to be changing at an alarming rate. And we reveal why the map of the world you've been using all your life is a complete and utter fabrication. Read on.

Climate change isn't the only serious environmental threat facing the planet. There are also crises unfolding over plastic pollution in the oceans, deforestation, overpopulation, and food production, to name just a few. Suffice it to say, planet Earth is in something of a pickle, and it's going to take some smart minds and a concerted effort on the part of humanity to put things right.

But the good news, at least for those of us who value our sanity, is that it's not all doom and gloom. A study by historians from Harvard has found that—despite climate change, and the general crumbling of the world order—now is actually a pretty good time to be alive. Things were markedly worse around the sixth century CE, when a volcanic eruption is believed to have thrown ash into the atmosphere, blocking out the Sun and causing crop failures, widespread famine, death, misery, and the fall of empires.

So cheer the fuck up.

ENVIRONMENT

MAN WHO FELL INTO YELLOWSTONE HOT SPRING COMPLETELY DISSOLVED WITHIN A DAY

by Robin Andrews

BENEATH YELLOWSTONE NATIONAL PARK RESIDES ONE of the largest magma chambers in the world. Thanks to this unfathomably hot fuel source, the water systems around the park can often be extremely hot and acidic.

You definitely shouldn't take a dip in them. They will kill you.

Back in June 2016, a 23-year-old man fell into one, and he died fairly quickly. Subsequently, thanks to a Freedom of Information Act request by a local TV network, more grisly details of the cause and the aftermath were revealed.

Apparently, the man was looking for a place to "hot pot," which describes the act of getting slightly singed in natural hot springs for no logical reason whatsoever. He leaned over to dip his forefinger in, in order to test the temperature of the water, when he slipped and fell.

The victim was found dead and drifting around the pool later that day, but officials could not quite reach him to drag him out. A thunderstorm promptly arrived and forced them to retreat for the night.

Returning the next day, they found that nothing of the man remained—except his wallet and his flip-flops.

In his incident report, Deputy Chief Ranger Lorant Veress pointed out that the waters were particularly hot and acidic that day. "In a very short order, there was a significant amount of dissolving," he was quoted as saying by *Time* magazine.

Although incidents like this are clearly quite tragic, they're also testament to the incredibly daft lengths people will go in order to show off, be "brave," or—in this case—have a very unique bath.

Yellowstone's geothermal ponds, pools, and geysers average around 199°F (93°C) at the surface, and they are far hotter just a few yards (meters) down. The temperature of the liquid can exceed 212°F (100°C)—the boiling point—because the pressure of the overlying water prevents it from turning to steam.

Only inhabitable by a specialized group of organisms known as archaea, these watery doom portals are fenced off and surrounded by a bunch of quite prominent warning signs for a really, really good reason.

Are you a microscopic, extremophilic life-form? No, we didn't think so. So stay the hell back, and don't try any of this "hot potting" nonsense unless you want to dissolve like a sugar cube in coffee.

ENVIRONMENT

POMPEII SKELETON REVEALS THE "UNLUCKIEST GUY IN HISTORY"

by James Felton

A man who fled Pompeii midway through the eruption of Mount Vesuvius in the year 79 CE, only to be smashed to death by a massive falling slab of stone suffered one final humiliation in 2018: by becoming an internet meme.

The skeleton (perhaps, unsurprisingly, not the head) of the "unlucky" man was unearthed recently at Pompeii's archaeological dig site. It would appear that, while attempting to flee the eruption of Vesuvius, he had his head crushed by a gigantic, falling block of masonry. Since that day in 79 CE, he has remained trapped there with his face underneath the rock.

After CNN tweeted the image, with a short explanation of the story, the internet had a field day. Responses ranged from sympathy for the unlucky guy to outright laughter. Meanwhile, the story was retweeted over 14,000 times.

PHOTOGRAPHER CAPTURES AMAZING IMAGES OF WEIRDLY ALIEN "LIGHT PILLARS" FLOATING IN THE SKY

by Jonathan O'Callaghan

A PHOTOGRAPHER HAS TAKEN SOME RATHER AMAZING photos of an unusual phenomenon known as "light pillars."

Vincent Brady from Charlotte, Michigan, snapped the images above nearby Whitefish Bay, which is on the eastern edge of Lake Superior.

Light pillars can look almost alien-like, with vertical beams of light seeming to extend up into space. They have an earthly explanation, though—they're the result of cold air carrying flat, plate-like ice crystals, which reflect artificial lights to produce the odd effect.

"Light pillars have become one of my favorite subjects," Brady told IFLScience. "They're a unique atmospheric phenomenon that occurs when the temperature drops to single digits, but more likely when it's subzero and the wind is calm.

"As moisture rolls in, it becomes crystallized, often referred to as 'diamond dust.' You can see it glistening in lights, seemingly defying gravity and dancing through the air. As the ice crystals float over light sources, they reflect and refract light and produce what appear as light pillars. Seeing them is a visual treat."

The crystals also sometimes melt as they approach the ground, which can give the pillars the added effect of appearing to hover in the air. "You can think you're seeing some sort of alien invasion," said Brady.

Interestingly, it's not just artificial lights that can produce light pillars. The Sun can do it too, an effect known as—you guessed it—a Sun pillar. The exact same thing is happening, as the Sun's light reflects off the ice crystals to create a glowing vertical shard. The Moon can also produce a similar effect (we'll let you work out the name).

Sun pillars are best viewed when the Sun is low in the sky. Light pillars, meanwhile, can be seen throughout the night when there are plenty of artificial lights on the ground to illuminate the crystals.

You can see more of Brady's work at his website (www.vincentbrady.com).

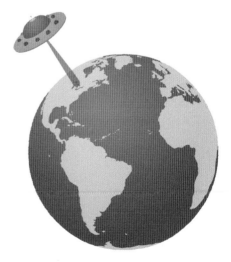

ANTARCTICA IS NOW MELTING SIX TIMES FASTER THAN IT WAS IN 1979

by Rachel Baxter

ANTARCTICA IS MELTING AT AN UNPRECEDENTED RATE. Between 1979 and 2017, Antarctic ice loss increased by a factor of six, causing sea levels to rise by half an inch (13 mm). That's according to a study published in 2019 in the *Proceedings of the National Academy of Sciences*.

An international team of scientists from the University of California, Irvine (UCI), NASA's Jet Propulsion Laboratory (JPL), and Utrecht University, in the Netherlands, conducted the "longest-ever assessment of remaining Antarctic ice mass." The team looked at aerial and satellite images of 18 Antarctic regions, which included 176 basins and some surrounding islands, to see how they had changed over the past four decades.

They discovered that, from 1979 to 1990, Antarctica lost about 44 billion tons (40 billion tonnes) of ice each year. After a slow rise between 1979 and 2001, the rate of ice loss suddenly jumped by 280 percent to 148 billion tons (134 billion tonnes), reaching an unthinkable 278 billion tons (252 billion tonnes) between 2009 and 2017.

This ice loss contributes to sea-level rise, and the team found that Antarctica's melting ice caused sea levels around the world to rise by 0.5 inches (13 mm) during the decades studied.

"That's just the tip of the iceberg, so to speak," said lead author Eric Rignot, a professor at UCI and senior project scientist at JPL, in a statement. "As the Antarctic ice sheet continues to melt away, we expect multi-meter sea-level rise from Antarctica in the coming centuries."

And Antarctica isn't the only contributor to sea-level rise: A recent study found that our oceans are warming at a faster rate than expected due to climate change, and, thanks to thermal expansion, warmer waters mean rising seas. This, in turn, threatens coastal communities as flooding becomes more extreme.

Somewhat unexpectedly, the researchers also found that the eastern Antarctic is an important contributor to ice loss, more so than previously thought. Past studies had suggested little to no loss of ice from this part of the continent.

"The Wilkes Land sector of East Antarctica has, overall, always been an important participant in the mass loss, even as far back as the 1980s, as our research has shown," said Rignot. "This region is probably more sensitive to climate [change] than has traditionally been assumed, and that's important to know, because it holds even more ice than West Antarctica and the Antarctic Peninsula together."

It's no secret that the frozen continent is melting more rapidly now due to human-induced global warming. To prevent climate catastrophe, we urgently need to reduce our greenhouse gas emissions by switching to renewable, non-polluting fuels.

WHEN WAS THE WORST TIME TO BE ALIVE IN HUMAN HISTORY?

by Tom Hale

BELIEVE IT OR NOT, WE CURRENTLY LIVE IN WHAT IS the safest era of human history. Sure, "strongman politics" has made a comeback, many of the planet's biggest problems remain unsolved, and there was that god-awful year when half of the world's most beloved celebrities dropped dead. Nevertheless, relatively speaking, now is a great time to be alive.

So, when was the crappiest time to be alive? This question was inadvertently raised by a historical study carried out in 2018, attempting to figure out how the European monetary system changed after the fall of the Western Roman Empire. Writing in the journal *Antiquity*, the researchers were looking for evidence of pollution from silver processing in ice cores buried deep in the European Alps. In doing so, they came across all kinds of insights into natural disasters and climate change events through the centuries.

One thing was clear: The century following the year 536 CE was a goddamn miserable time for all concerned.

"It was the beginning of one of the worst periods to be alive, if not the worst year," study author Michael McCormick, a medieval historian at Harvard, told *Science* magazine.

This era was grim, not because of bloody wars or ferocious diseases, but due to a number of extreme weather events that led to a widespread famine. Although there are many theories floating around as to why the famine occurred, some of the sturdiest evidence points toward a "volcanic winter," where ash and dust were thrown into the air from the eruption of a volcano, thereby obscuring the Sun.

Nobody is completely certain which volcano was the culprit. El Salvador's Ilopango has long stood as a top contender. However, the 2018 study hints that the eruption took place in Iceland, as the ice cores gathered contain volcanic glass that's chemically similar to particles found across Europe and Greenland.

Whatever the volcano, its effects were widespread, sparking the "Late Antique Little Ice Age" and a chain of global crop failures and famine. Snow fell during the summer in China, and droughts hit Peru. Meanwhile, Gaelic Irish annals talk of "a failure of bread in the year 536 [CE]." It seems there was scarcely a corner of Earth left unscathed.

The mini–ice age also raised a load of social problems. Some researchers have even argued that the effects of the volcanic event in 536 CE were so profound that they brought down empires (or at least tipped them over the edge). As noted in a 2016 study in *Nature Geoscience*, the century after the volcanic eruption saw the collapse of the Sasanian Empire, the decline of the Eastern Roman Empire, political upheavals in China, and many other instances of bloody social turmoil across Eurasia.

All in all, a crummy time to be alive.

72

Organic Food Is Worse for the Climate Than Non-Organic Food

by James Felton

A study of Swedish crop farming, published in *Nature*, has found that organic peas have a 50 percent bigger impact on the climate than peas farmed conventionally. For Swedish winter wheat, the figure was closer to 70 percent. Fertilizer-free organic crops produce lower yields, meaning more land area must be used to cultivate them.

73

Scientists Have Spotted a "Lost Continent" Using Satellites

by Tom Hale

Lurking deep beneath the ice sheets of Antarctica, scientists have detected the remnants of long-lost continents. Writing in the journal *Scientific Reports*, the team explained their discovery of the ancient continents using gravity-mapping data gathered by the European Space Agency's Gravity Field and Steady-State Ocean Circulation Explore (GOCE) satellite.

ENVIRONMENT

HUGE 210-FOOT (64-METER) FATBERG DISCOVERED BENEATH QUAINT ENGLISH SEASIDE TOWN

by Rosie McCall

EVER SINCE LONDON'S MOST UNLIKELY CELEBRITY, THE Whitechapel fatberg, drew headlines in 2017, the UK has had something of a morbid fascination with the mammoth-sized clumps of gunk lurking in the sewer system.

The latest fatberg to make the rounds is a 210-foot (64-meter) beast found in Sidmouth, Devon—the largest of its kind to be found by the region's local water provider, South West Water (SWW). To put it into perspective, this particular monstrosity is about 26 feet (8 meters) longer than the Leaning Tower of Pisa is tall.

Fortunately for residents of and visitors to the sea-facing town of Sidmouth, authorities located the berg "in good time" and long before it posed any kind of risk to beachgoers or risked blocking any toilets, BBC News reports.

Fatbergs are an abominable concoction of human excrement, sanitary products, drugs, contraceptives, and other insoluble items, all covered in congealed fat. As IFLScience reported in 2017, Thames Water spends more than $1 million every single month to remove the swarm of fatbergs plaguing the UK capital, thanks to residents' willful disregard for proper waste management practice.

Fortunately for the brave men and women at SWW, this fatberg didn't weigh in anywhere close to the Whitechapel monstrosity. Cheerfully nicknamed "Fatty McFatberg," this weighed 143 tons (130 tonnes) and was a jaw-dropping 820 feet (250 meters) tip to tip. It was so big it took a dedicated team of eight, working every day for several weeks, to break it down with the help of some high-powered jet hoses.

And even that beast of a fatberg pales in comparison to one found close to the South Bank in central London in April 2018, a 2,460-foot-long (750-meter-long) monster. That's roughly twice the height of the Empire State Building.

The latest discovery shows that fatbergs are a much larger problem that we thought—and even country dwellers aren't safe from the lavatorial horrors that plague urban centers.

The moral of the story—think before you flush.

Microplastics Found in 100 Percent of Sea Turtles Tested

by Tom Hale

A collaboration of UK scientists conducting a study of 102 turtles, across seven different species, from three different oceans, has found that all individuals—every single one—had microplastics in their guts. In total, over 800 synthetic particles were discovered in the animals' digestive tracts.

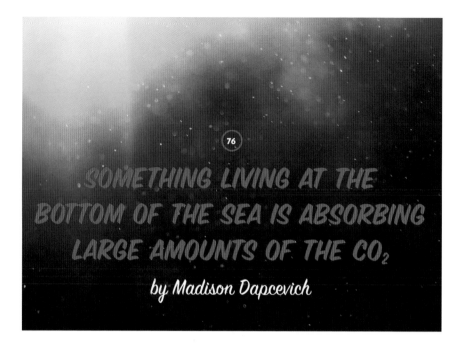

SOMETHING LIVING AT THE BOTTOM OF THE SEA IS ABSORBING LARGE AMOUNTS OF THE CO₂

by Madison Dapcevich

BACTERIA LIVING MORE THAN 2½ MILES (4 KM) BELOW the surface of the Pacific Ocean are absorbing an estimated 10 percent of the carbon dioxide that oceans remove from the atmosphere every year.

A team of researchers discovered that benthic bacteria (those living just above the seafloor) are taking up large amounts of CO_2 and assimilating it into their biomass by an unknown process. "This was completely unexpected," said study author Andrew Sweetman in a statement. "Their biomass then potentially becomes a food source for other animals in the deep sea, so actually what we've discovered is a potential alternative food source in the deepest parts of the ocean, where we thought there was none."

Writing in *Limnology and Oceanography*, the researchers say benthic bacteria, rather than seafloor animals, could be the "most important organisms" when it comes to consuming organic waste that drifts down toward the ocean floor.

To examine the cellular processes of benthic organisms, the team analyzed sediment samples taken from an area in the eastern Pacific Ocean between Hawaii and Mexico, known as the Clarion-Clipperton Fracture Zone (CCFZ), a deep-sea ecosystem completely devoid of light, but for flashes of bioluminescence, and with a surprisingly biodiverse seabed environment. Bacteria here "dominated the consumption" of organic waste over a period of one or two days for which the team observed. When scaling their results, that equates to about 220 million tons (200 million tonnes) of carbon dioxide that could be fixed into biomass every year, making the region a potentially important component in the deep-sea carbon cycle.

"We found the same activity at multiple study sites separated by hundreds of kilometers, so we can reasonably assume this is happening on the seabed in the eastern CCFZ and possibly across the entire CCFZ," said Sweetman.

Assuming the results can be applied to the greater CCFZ, the authors say their findings could have implications for proposed mineral extraction in this region. The CCFZ is home to more than just deep-sea sponges, sea anemones, shrimps, and octopods. The clay-like muddy bottom is topped with trillions of potato-sized polymetallic nodules containing deposits of nickel, manganese, copper, zinc, cobalt, and other minerals. It's an area so rich in minerals that the International Seabed Authority has awarded 16 exploration contracts for groups interested in conducting surveys for future seabed mining.

"If mining proceeds in the CCFZ, it will significantly disturb the seafloor environment," said Sweetman. "Just four experiments similar to ours have been conducted in situ in the abyssal regions of the oceans; we need to know much more about abyssal seafloor biology and ecology before we even consider mining the region."

A MYSTERY ABOUT EASTER ISLAND'S STATUES MIGHT FINALLY BE SOLVED

by Tom Hale

A SELECT FEW OF EASTER ISLAND'S GIANT "MOAI" HEADS have an especially strange choice of headwear. Just a handful of these famous volcanic-rock dudes have a red "hat" placed on top of their head. Not only are these hats made out of a red stone from a quarry on the other side of the island, they also appear to have been added after the statues were erected.

ENVIRONMENT

So, how and why did these 11-ton (10-tonne) "pukao" hats get up there on top of a 13-foot (4-meter) statue? After years of contemplation, archaeologists from the Pennsylvania State University think they've finally found the answer.

"Lots of people have come up with ideas, but we are the first to come up with an idea that uses archaeological evidence," Sean W. Hixon, a graduate student in anthropology at Penn State, said in a statement.

The island, 2,340 miles (3,700 km) off the coast of Chile, was first inhabited in the 13th century by Polynesian seafarers. They spent the next two centuries forging these giant sculptures out of volcanic tuff. Experts still argue about it, but it's now generally accepted that these monoliths were moved from the quarry with a rocking motion, in much the same way that you would move a refrigerator, except using ropes. This quarry, however, is over 7 1/2 miles (12 km) away from the red scoria quarry that the pukao hats were cut from.

Writing in the *Journal of Archaeological Science*, Hixon and colleagues suggest that the hats were placed on the statues after they were erected, most likely after being rolled in raw form from the quarry to the statues and then carved on site. They argue this because uncarved red scoria stone can be found en route to the statues from the quarries.

Ramps must have then been used to get the hats up to head level. However, imaging studies have highlighted indentations on the under sides of the hats, which would have rubbed off the soft stones had they each been pushed up a ramp on their base.

Instead, the researchers argue the pukao were rolled up the ramps to the top of a standing statue using a parbuckling technique, a tried-and-tested method of shifting heavy loads that employs the help of ropes looped under the object (see illustration on page 148-149). The researchers figured out that this whole process would require fewer than 15 workers.

Why they went to all this effort, though, still remains a mystery.

An Island off the Coast of Japan Has Gone Missing

by Rosie McCall

The island Esanbe Hanakita Kojima (population: zero), off the north coast of Hokkaido, Japan, has gone missing. The island, 1,640 feet (500 meters) off the coast of the village of Sarufutsu, stood at just 5 feet (1.4 meters) above sea level. It is thought to have now become submerged.

ENVIRONMENT

THE MAP YOU GREW UP WITH IS A LIE. THIS IS WHAT THE WORLD REALLY LOOKS LIKE

by Tom Hale

THE WORLD IS NOT QUITE WHAT IT SEEMS. MOST OF the maps you see are based on "Mercator projection," a 450-year-old technique for projecting the spherical surface of the Earth onto a flat sheet. Even Google Maps used a variant of it until August 2018.

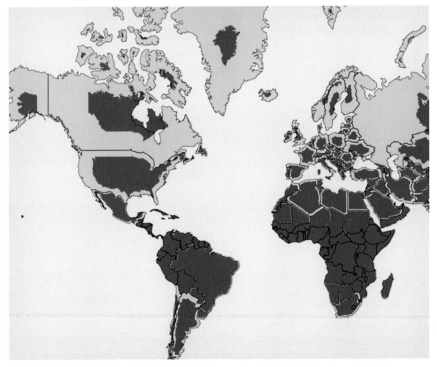

The Mercator projection (light blue) overestimates the true size of landmasses (dark blue) the further you get from the equator.

However, Mercator projection is surprisingly bad at accurately reflecting the true size of many countries. To demonstrate, Neil Kaye, a climate data scientist at the Met Office in the UK, has designed a map visualization that compares the Mercator projection with true projections of each country's land area. The image (see page 151) elegantly shows how the Mercator projection overrepresents the size of many countries, especially those farther away from the equator.

Just take a look at how much smaller Russia, Canada, and Greenland really are. Europe, parts of Asia, and the US also shrink away a considerable amount using the new projection. The old Mercator projection depicts Greenland as a landmass larger than Africa, however, in reality, Africa's area is a whopping 14 times greater than Greenland's.

The Mercator projection was first presented by the Flemish cartographer Gerardus Mercator in 1569. It's been extremely useful for exploration, as it allows a navigator to plot a straight-line course and maintains the country's true shape, but when you translate a three-dimensional shape, such as a globe, into a two-dimensional projection, something's got to give. In this case, it's a distortion of size and distance as you get closer to the planet's poles.

The Mercator projection has also been accused of having political undertones by presenting a Eurocentric colonial view of the world. As a result of all of these biases and shortcomings, some schools in Boston even decided to get rid of the map in favor of the alternative Gall-Peters world map. However, this projection isn't perfect, either. While it represents the area of a landmass more accurately, it distorts the shape.

In August 2018, cartographers released the alternative Equal Earth projection map in hopes of overcoming all the problems with the world's various different map projection schemes. However, if past debates within cartography (of which there have been many) are anything to go by, then this issue will undoubtedly rumble on and on.

ENVIRONMENT

2015–2018 Have Been the Hottest Years on Record, UN Report Reveals

by Tom Hale

The United Nations' World Meteorological Organisation (WMO) has announced that 2015, 2016, 2017, and 2018 are the top four hottest years ever recorded. The WMO warned that the current trend would lead to temperature increases of 3–5°C by the end of the century. The message on climate change couldn't be clearer.

Mass Grave of Child Human Sacrifice Victims Found in Peru

by Jonathan O'Callaghan

A report by *National Geographic* has found the largest incidence of mass child sacrifice in the Americas, and possibly the world. The discovery was made at Huanchaquito–Las Llamas, on Peru's northern coast. In total, they found remains of more than 140 children, thought to date from between 1400 and 1450.

IFLSCIENCE

(82)

EARTH'S MAGNETIC FIELD IS UP TO SOME SERIOUSLY WEIRD STUFF AND NO ONE KNOWS WHY

by Tom Hale

THE PLANET'S MAGNETIC FIELD IS UP TO MISCHIEF again and geophysicists are pretty dumbfounded.

Earth's magnetic poles can wander several kilometers every year; however, the north pole's movement has become increasingly stranger of late. For reasons that are currently unclear, the magnetic north pole seems to be slipping away from Canada and toward Siberia at an erratic rate, according to a news report published in *Nature* early in 2019.

"The location of the north magnetic pole appears to be governed by two large-scale patches of magnetic field, one beneath Canada and one beneath Siberia," Phil Livermore, a geomagnetist at the University of Leeds, in the UK, was quoted as saying by *Nature*. "The Siberian patch is winning the competition."

Every five years, the National Oceanic and Atmospheric Administration (NOAA) maps out the Earth's magnetic field in the World Magnetic Model (WMM). At the time of writing (January 2019), this had last been published in 2015, with the next edition planned for 2020. But the freak magnetic behavior has forced scientists to revise the map earlier than anticipated (although that had to be postponed because of the US government shutdown).

Earth's magnetic field is created by molten iron in its core swirling around because of thermal convection currents. It's a fairly chaotic situation in there, resulting in a complex pattern of magnetism, which can prove extremely difficult to model and predict. Just to complicate things further, an unusually punchy geomagnetic pulse (a sharp change in the strength of the Earth's magnetic field) occurred under South America in 2016, which is believed to have contributed to the recent unexpected changes.

The magnetic field is central to many forms of navigation. Most obviously, a compass relies on magnetic fields, but more advanced systems use the field to get a bearing, too. Perhaps more worryingly, it is the magnetic field that protects the Earth, and all life on the planet's surface, from lethal, high-energy, charged particles that stream through outer space.

It is possible for truly monstrous changes to the magnetic field to take place. Scientists know the Earth can undergo a phenomenon known as a "geomagnetic reversal"—where north and south magnetic poles literally swap places, with the field becoming significantly weakened during the switchover. The last major reversal was 781,000 years ago, but smaller flips are believed to have occurred every 20,000–30,000 years over the last 20 million years.

NEW RESEARCH SUGGESTS ITALIAN SUPERVOLCANO IS FILLING UP WITH MAGMA

by Alfredo Carpineti

WHEN ONE THINKS OF NAPLES AND VOLCANOES, THE mind goes straight to Mount Vesuvius, the volcano responsible for the destruction of Pompeii and Herculaneum in 79 CE. But a more worrying volcano exists just to the west of the city, called Campi Flegrei. Also referred to as the Phlegraean Fields, this "supervolcano" has a large crater (aka caldera) that's 8 miles (13 km) across. An eruption would be deadly and catastrophic, spewing magma tens of kilometers into the stratosphere.

Research published in 2018, in the journal *Science Advances*, has shown that the magma reservoir underneath Campi Flegrei has begun to fill up. This buildup phase will likely lead to a large-volume eruption in the distant future, and while the danger is not imminent, it shows that the supervolcano needs to be kept under constant surveillance. About 1.5 million people live next to or right on top of the caldera.

The new assessment shows that the chemical composition of the magma entering the caldera has changed lately. Volatile substances, in particular, are being separated from the magma, which is increasing the pressure beneath the caldera.

The two major past eruptions at this site occurred 39,000 and 15,000 years ago, which formed the caldera and led to part of it ending up underwater. Several minor eruptions also contributed to the land-

ENVIRONMENT

scape of the region, and the researchers were curious to see what they could learn about the future of the supervolcano from its recent past. For instance, in 1538, an eight-day eruption led to the formation of Monte Nuovo (literally "new mountain"), a cinder-cone volcano, within the Campi Flegrei caldera.

While the analysis gives important insights into the changes within Campi Flegrei, the team states clearly that there is no evidence on which to base speculation about when the next eruption will be, or if it will be a major one. There is even the possibility that the system will become dormant.

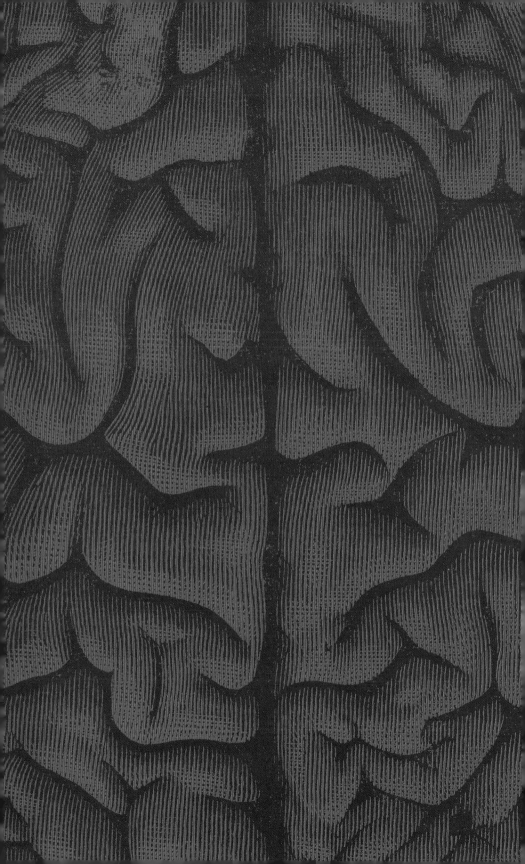

CHAPTER SIX

BRAIN

For a 3.3-pound (1.5-kg) fatty blob with the approximate consistency of tofu, the human brain has to be admired. After all, it's managed to formulate the theories of relativity and quantum mechanics, written all the great novels, and, perhaps most astoundingly of all, has evolved sentience, self-awareness, and the ability to contemplate its own origin and existence.

That said, it's also given us John Travolta's *Battlefield Earth* and the truly execrable *Dirty Grandpa*.

For the most part, though, the human brain is a veritable powerhouse. It generates around 23 watts of heat while its owner is awake, and consumes around 20 percent of the oxygen that's funneled into the bloodstream from the lungs. The brain of an adult human is made up of around 100 billion nerve cells, or neurons. These communicate with each other through electrical impulses and by the exchange of messenger chemicals, known as neurotransmitters. Each neuron can be wired like this to as many as 10,000 others, forming a network of up to 1,000 trillion connections. And the structure of these connections is responsible for storing our knowledge and memories.

Perhaps the biggest outstanding mystery of the brain is consciousness. How did this tangled web of synaptic connections develop the power not just to store information and perform computations, but to be aware and to have conscious experiences? These experiences—the redness of the color red, the flavor of a burrito, the smell of a sea breeze—are known as "qualia" to neuroscientists, and they are subjective to each individual, defying description except perhaps by analogy.

Qualia must correspond to certain physical processes in the brain, and identifying these so-called "neural correlates of consciousness"—the firing of a certain group of neurons to signify the aroma of a rose, for example— has been a major focus for scientists.

In 2014, a breakthrough was made when researchers found that electrically stimulating a brain region called the claustrum induced a loss of consciousness, which was immediately reversed the moment the current was switched off again. Later, in 2018, scientists were able to identify two parts of the brain that contribute to our experience of free will—our sense of being in control of our actions. You can read more about that discovery later on in this chapter. Other researchers have even developed theories positing that free will and our experience of qualia all derive from applying the weird laws of quantum mechanics to the functioning of the brain. However, these claims have been greeted with a good deal of skepticism.

In separate research, some astonishing insights into the nature of the brain have been revealed using medical scanners. Scientists have used functional magnetic resonance imaging (fMRI) to identify patterns of activity that they say can reveal whether someone's telling lies or being truthful, and can give away personal prejudices, political leanings, and even brand preferences (Coke or Pepsi?). FMRI works by coaxing radiation from hydrogen atoms residing within water molecules in the body by stimulating them with radio waves and magnetic fields. Measuring the strength of the radio waves emitted in response reveals regions of increased blood flow. And this acts as a tracer for which brain regions are most active.

More recently, fMRI scanning has been combined with a field called "machine learning." This is a branch of data science that enables computers to learn and make discoveries from raw data, doing so with minimal human intervention. Researchers at Kyoto University in Japan gathered fMRI scans of patients' brains, taken while they looked at various images. The set of scans and the input images were fed to a kind of machine-learning agent called a "deep neural network," which was then able to look at new fMRI scans and deduce with reasonable accuracy what pictures the subjects in each case were thinking about. Pretty cool, huh?

More on that later. Brains scans can even yield insights into the decision-making process. This was demonstrated by scientists who were able to peer inside the brains of rats. In an experiment that you'll soon read about, the researchers placed rats in a maze and, based on the observed patterns of brain activity, they were able to predict which path through the maze each animal was about to take.

Also coming up, find out how magic mushrooms can boost your creativity. Researchers from Leiden University, in the Netherlands (surprise, surprise), have found that so-called microdosing (taking tiny amounts of psilocybin, the active ingredient in 'shrooms) led test subjects to come up with more original solutions to problems. (And, just to be clear, no, none of these involved enlisting the help of ninja unicorns on mopeds.)

Plus, we discover the facial feature that's a dead giveaway of a narcissist. Find out just why it is that one or two (or was it 10?) drinks too many can leave you a little hazy on the details of what went on last night. And get the lowdown on why conscientiousness is, apparently, the most attractive characteristic in a sexual partner (according to heterosexual couples, anyway).

As Miles Monroe put it in the 1973 movie *Sleeper*: "My brain? It's my second favorite organ!"

WE JUST FOUND THE PART OF THE BRAIN RESPONSIBLE FOR FREE WILL
by Rosie McCall

THE PHILOSOPHER THOMAS HOBBES DESCRIBED FREE will (or "liberty") as "the absence of all the impediments to action that are not contained in the nature and intrinsical quality of the agent"—which, in plain English, is the ability to act without any outside constraint, be that an overbearing boyfriend/girlfriend or something altogether more whimsical, like fate.

The scientific definition, however, is far more specific. Essentially, it comes down to two cognitive processes: volition and agency. Volition is "the desire to act," whereas agency is "the sense of responsibility for our actions."

Thanks to a study published in the *Proceedings of the National Academy of Sciences* in 2018, scientists have identified the exact locations in the brain responsible for these processes and, therefore, our perception of free will.

Researchers from Beth Israel Deaconess Medical Center, in Boston, studied 28 cases where brain injury had affected patients' volition and left them with no desire to move or speak (a condition known as "akinetic mutism"). They also examined a further 50 cases where agency had been impaired, so that the patients were left feeling as though their movements weren't their own ("alien limb syndrome"). The team looked at scans to find the lesions in each patient's brain responsible.

This revealed a diverse range of injury locations but, interestingly, all lesions appeared within one of two networks. The injuries related to akinetic mutism were connected to the brain's anterior cingulate cortex (responsible for motivation and planning), whereas almost all the injuries (90 percent) related to alien limb syndrome were connected to the brain's precuneus cortex.

To confirm that these two networks are indeed the two areas responsible for free will (as defined by science, not Hobbes), the researchers examined the effect of brain stimulation on free-will perception in healthy volunteers and looked at images of the brains of psychiatric patients with abnormal free-will perception. Both revealed alterations in the same networks.

While the information is not going to stop moral philosophers and sociologists from debating the meaning (and the very existence) of free will, the researchers hope it will prove useful when it comes to helping patients with volition- and agency-inhibiting injuries.

BRAIN

ARTIFICIAL INTELLIGENCE RE-CREATES IMAGES FROM INSIDE THE HUMAN BRAIN

by Jonathan O'Callaghan

A TEAM OF RESEARCHERS SAY THEY HAVE USED MACHINE-learning to re-create images from "thoughts" inside patients' brains.

The research, reported in January 2018 by *Science* magazine, was conducted by scientists from Kyoto University in Japan and led by Yukiyasu Kamitani. Using functional magnetic resonance imaging (fMRI), the team said they were able to reconstruct images by analyzing scans of brain activity. An artificial intelligence agent, known as a deep neural network (DNN), was able to process the scans to re-create pixel-by-pixel images that resembled the originals.

"The results suggest that hierarchical visual information in the brain can be effectively combined to reconstruct perceptual and subjective images," the team wrote in their paper.

The research builds on earlier work by the same team that found that brain activity could be decoded to reveal the sensory inputs producing it. Other researchers have reported similar work in this field.

"This is a significant improvement on their earlier work," Professor Geraint Rees, a neuroimaging expert from University College London, told *The Times*.

In this latest paper, the researchers used three subjects (two males, aged 33 and 23, and one female, aged 23). They were shown images of things like a mailbox and a lion, as well as geometric shapes and alphabetical letters.

The subjects viewed the images while inside an fMRI scanner, with their heads held securely in place. They took part in multiple scanning sessions, each lasting a maximum of 2 hours, spread over a period of 10 months.

The participants first stared at each image for a number of seconds, while their brain activity was recorded. This data, alongside the input image, was then used to train the DNN—to teach it the patterns of brain activity corresponding to different features in a visual image.

Later, the subjects were asked to remember one of the images they had seen and picture it in their mind. Using the DNN, the researchers then attempted to decode the signals recorded by the fMRI scanner and produce a computer-generated reconstruction of the original image.

Some of the results were rather remarkable, with the DNN able to reproduce images of a DVD player, feet with socks on, a fly, and more. However, it wasn't too hot on other images—like a person with a cowboy hat, or a snowmobile.

"Our approach could provide a unique window into our internal world by translating brain activity into images," the team noted.

Here's a US Army Trick for Falling Asleep Anywhere in 120 Seconds

by Jonathan O'Callaghan

Relax your facial muscles, drop your shoulders, and relax your arms. Breathe out, relax your chest, and then your legs. Blank your mind. Repeating the words *Don't think* may help. After US Navy pilot trainees practiced this for six weeks, 96 percent could fall asleep in under two minutes.

GROWING UP POOR PHYSICALLY CHANGES THE STRUCTURE OF A CHILD'S BRAIN

by Madison Dapcevich

A LONG-TERM ANALYSIS OF HUNDREDS OF ADOLESCENT brains suggests that the socioeconomic status (SES) of a child's family may play a role in the early development of brain areas responsible for learning, language, and emotion.

To study the connection between parents' income and education levels and their children's cognitive development, researchers from the US National Institute of Mental Health scanned the brains of more than 600 individuals over the course of their lives between the ages of 5 and 25. Then they compared these neuroimages against data on their parents' education and occupation, as well as each participant's IQ.

When it comes to the relationship between brain anatomy and SES, not much changes from childhood to early adulthood. This led researchers to suspect that preschool life is a pivotal time in which associations between socioeconomic status and brain organization first begin to develop.

The authors, writing in the *Journal of Neuroscience*, found correlations between SES and total gray matter volume, and between volume levels in the prefrontal cortex, which is the area of the brain associated with personality development, and the emotion-regulating hippocampus. The areas of the brain responsible for emotional development, learning, and language skills were found to be more complex in youngsters whose parents were better educated and worked in professional careers.

"Early brain development occurs within the context of each child's experiences and environment, which vary significantly as a function of socioeconomic status," wrote the authors. A child's early life experiences and environment depend on her family background, through factors such as her parents' income, education, and occupation. And these impact a child's mental health, cognitive development, and her academic achievements. Understanding how such things physically change the brain could help researchers understand how SES is associated with different life outcomes.

The researchers note that the link they have found between SES and cognitive development represents only one possible set of interactions between childhood environment, anatomy, and cognition.

People Would Rather Save a Cat Than a Criminal in Worldwide Trolley Problem Study
by Jonathan O'Callaghan

In an internet survey, scientists from the Massachusetts Institute of Technology posed a hypothetical moral dilemma, asking participants whether they thought a car with a family in it should swerve to save the lives of a criminal, a dog, or a cat. Dogs were saved the least, followed by criminals, and then cats.

A TECHNIQUE TO CONTROL YOUR DREAMS HAS BEEN VERIFIED FOR THE FIRST TIME

by Stephen Luntz

GET READY TO HAVE THE DREAMS OF . . . WELL, OF YOUR dreams. A technique to induce lucid dreaming—a state of dreaming that can be consciously experienced and controlled—was independently verified for the first time in 2017. More than half the participants in the study lucidly dreamed during the trial, a record-breaking success rate.

Lucid dreaming is the term given to the state where dreamers are aware

that they are dreaming, and have some control over how the dream progresses. Once considered a myth, science has confirmed that lucid dreams exist, and has found techniques to help induce them. Nevertheless, some of these require advanced equipment, and others are far from reliable.

This is unfortunate, as they are considered a potential tool for healing traumas and for controlling unhealthy behavior. So Dr. Denholm Aspy, of the University of Adelaide, decided to investigate.

Aspy instructed 169 participants in techniques developed to induce lucid dreaming. One of these, called "reality testing," gets people into the habit of regularly checking to make sure they really are awake. The other, called "mnemonic induction of lucid dreams" (or MILD), has participants set alarms to wake them after five hours and recite the line "The next time I am dreaming, I will remember that I'm dreaming," before going back to sleep.

Writing in the journal *Dreaming* in 2017, Aspy reports that reality testing on its own produced no benefit, but of those who tried the combination of reality testing and MILD, 53 percent had a lucid dream during the trial period, while 17 percent were successful each night. This exceeds the results of any previous study conducted without physical interventions, such as trying to awaken participants once they were in dream sleep, he told IFLScience.

Approximately 55 percent of us experience a lucid dream at some point in our life. Aspy himself became interested in lucid dreaming after having one as a child, and says he made it the topic of his psychology PhD research after having a lucid dream the night before he was due to begin his doctoral studies.

Most lucid dreamers initially wake up quickly, Aspy told IFLScience, but with practice they can learn to extend their lucid dreams for up to an hour.

HERE'S WHAT HAPPENS TO ALCOHOLICS' BRAINS WHEN THEY QUIT DRINKING

by Ben Taub

SCIENTISTS HAVE COME UP WITH SURPRISING EVIDENCE that may explain why recovering alcoholics find it so hard to stay off the booze.

Publishing their findings in 2016 in the *Proceedings of the National Academy of Sciences,* the researchers suggest that when an alcoholic stops drinking, the brain's ability to use the neurotransmitter dopamine changes, altering the wiring of the brain's reward system.

Like many drugs, alcohol is known to stimulate the production of the chemical messenger dopamine, which activates the so-called reward

center of the brain. Previous studies into the nature of addiction have revealed that the dopamine response is significantly reduced in alcoholics, leading to a need to drink more in order to feel a buzz. But why?

To investigate, researchers began by examining brain tissue from deceased alcoholics. They found that these brains had fewer D1 dopamine receptors than normal brains. D1 receptors are the sites on the membranes of neuronal cells to which dopamine binds, causing these neurons to become excited. Any reduction in these receptor sites would therefore be expected to decrease the brain's responsiveness to dopamine, explaining why alcohol fails to satisfy.

The brains were also found to have fewer dopamine transporter sites, which allow for any unused dopamine to be sucked back up and recycled. As with D1 receptors, the disappearance of these sites is likely to hinder the brain's ability to use dopamine.

Next, the study authors used radiography techniques to track dopamine levels in the brains of alcohol-dependent rats that were denied alcohol for several weeks.

They discovered that dopamine levels dropped during the first six days. However, after three weeks, the researchers noted that dopamine levels were in fact elevated, as the number of available receptor and transporter sites plummeted, so that the rats' brains resembled those of the deceased alcoholic humans. Significantly, at the three-week mark, the rats displayed continued behavioral effects associated with alcohol cravings.

The study authors concluded that, while acute alcohol withdrawal may be associated with lowered dopamine levels, prolonged abstinence actually leads to dopamine levels in the brain being higher than normal. Crucially, they say that both of these states are representative of a dysfunctional reward system, increasing an addict's vulnerability to relapse.

PINK ISN'T REAL

by Tom Hale

Forget flamingos, cherry blossoms, and bundles of cotton candy. We hate to break it to you, but pink isn't a real color, at least not in the way you might think.

Visible light—red, orange, yellow, green, blue, indigo, and violet—is the chunk of frequencies in the electromagnetic spectrum that our eyes are able to perceive. But unlike most colors we come across, we cannot represent pink with a single frequency of light. So does this mean that pink isn't really a color?

"Of course pink is a color," Jill Morton, an expert in color theory and color psychology, told *Popular Science* in 2012. "But with that said, pink is indeed not part of the light spectrum. It's an extra-spectral color, and it has to be mixed to generate it."

If you're sticking to the rules of the electromagnetic spectrum, then, it might be more accurate to call pink "a tint of red."

THE KEY TO A HAPPY SEX LIFE SOUNDS PRETTY UNSEXY, ACCORDING TO THIS STUDY

by Tom Hale

HOLLYWOOD MOVIES AND OTHER ROMANTICIZED DEPICTIONS of love like to imagine "good sex" as a spontaneous splurge of impulsive animal desire. However, according to a new German study, the key to a healthy and fulfilling sex life could actually be . . . conscientiousness. Wild, right?

A study published in July 2018 in the *Journal of Sex Research* looked at what personality traits and partner types tend to create a happy sex life. They came to the unlikely conclusion that those who are big on planning and organization tend to report fewer problems and higher levels of satisfaction in the bedroom.

Psychotherapists from Ruhr University in Germany reached this conclusion by quizzing 964 German couples—98 percent of whom, it should be added, were in heterosexual relationships—about their sex life and satisfaction. They were asked intimate details, such as how easily they were aroused, how inhibited they were, and how they thought they performed sexually.

The researchers also asked the respondents to complete a questionnaire to assess how they scored on the Big Five personality traits: Conscientiousness, Agreeableness, Extroversion, Neuroticism, and Openness to Experience.

"Studies have shown that most of these personality traits and sexuality-related traits are relevant," Julia Velten, a post-doctoral fellow in clinical psychology and psychotherapy, told the news website *Quartz*.

However, it came as a surprise how strongly conscientiousness was correlated with sexual satisfaction. The researchers speculate that people who display this trait might actually make more thoughtful lovers, as they are more likely to carefully engage with their partners, make sure they are fulfilled, and remain focused on resolving any hiccups in the relationship.

"Individuals low on emotional stability or agreeableness may be more likely to behave in a way (i.e., express criticism, avoid communication) that triggers a negative response from a partner, which in turn may lead to inadequate sexual communication and result in lower sexual functioning," the study authors write.

"High conscientiousness can be especially beneficial when it comes to putting effort into a satisfying sexual life or to postponing one's own needs and interests to focus on resolving a sexual problem within the context of committed, long-term relationships."

So, if you're looking to spice things up in the bedroom, you just need to remember your pen and your day planner.

MICRODOSING MAGIC MUSHROOMS COULD SPARK CREATIVITY AND BOOST COGNITIVE SKILLS

by Tom Hale

ADVOCATES OF MICRODOSING CLAIM THAT TAKING TEENY doses of magic mushrooms and other psychedelic substances can inspire creative thought, boost your mood, and even enhance your cognitive function, all without the risk of a so-called "bad trip."

But aside from loose anecdotal evidence from Silicon Valley bros, what does the science say? A team of researchers from Leiden University in the Netherlands decided to find out. Their admittedly small-scale study, published in October 2018, is the first of its kind to experimentally investigate microdosing of magic mushrooms and its cognitive-enhancing effects within a natural setting.

Reporting in the journal *Psychopharmacology*, the researchers looked into how magic mushrooms, aka psilocybin or truffles, affected the brain function of 36 people at an event organized by the Psychedelic Society of the Netherlands. The participants were given a one-off dose of 0.01 ounces (0.37 grams) of dried truffles and asked to solve three puzzles.

It's worth noting that microdosing usually involves taking regular small doses in the hopes of obtaining a cumulative effect. Nevertheless, the researchers claim that they observed some subtle yet profound changes. The test subjects appeared to be drifting through the puzzle-solving tasks with great ease while creating solutions that were notably more original and flexible than those before they microdosed. This is what the study authors called "changes in fluid intelligence."

"Our results suggest that consuming a microdose of truffles allowed participants to create more out-of-the-box alternative solutions for a problem, thus providing preliminary support for the assumption that microdosing improves divergent thinking," lead author Luisa Prochazkova of Leiden University in the Netherlands explained in a statement.

"Moreover, we also observed an improvement in convergent thinking, that is, increased performance on a task that requires the convergence on one single correct or best solution." In sum, the findings of the study are pretty much what the anecdotal evidence has been hinting at for years.

The doors of scientific research into psychedelics have only recently opened, but there's been a wealth of research looking into their potential benefits. Some of the most promising findings so far have come from studies of their potential to ease depression and other mental health conditions. While the pros and cons are not yet crystal clear, many researchers are welcoming the fact that this intriguing subject is now open for critique and investigation.

"Apart from its benefits as a potential cognitive enhancement technique, microdosing could be further investigated for its therapeutic efficacy to help individuals who suffer from rigid thought patterns or behavior, such as individuals with depression or obsessive-compulsive disorder," Prochazkova explained.

Whether You're a Go-Getter or a Procrastinator Depends on This

by Tom Hale

Researchers from Ruhr-University Bochum, in Germany, say that whether you are a doer or a procrastinator depends on the way your brain is wired. They found people with poor action control, aka procrastinators, had brain scans indicating poor emotional control—suggesting doers are better at managing competing distractions and negative emotions.

95

HOW AND WHY ORGASM FACES DIFFER AROUND THE WORLD

by Tom Hale

ARE FACIAL EXPRESSIONS UNIVERSAL ACROSS THE GLOBE? You certainly might assume so. If you're hit on the thumb with a hammer, you'll make a pretty similar face to someone who was raised in a totally different culture on the other side of the planet.

Only, when it comes to orgasms, that's not true at all.

Research published in 2018 looked at how two different cultures, Western and East Asian, interpret facial expressions associated with pain and orgasms. Pain was pretty much universal. But they discovered some deep cross-cultural differences between orgasm faces. While Westerners expect to see their partner with a wide-eyed expression and a gaping mouth, East Asian people anticipate a smiling face with tight lips.

"This finding is counterintuitive, because facial expressions are widely considered to be a powerful tool for human social communication and interaction," the study authors wrote.

As reported in the *Proceedings of the Natural Academy of Sciences*, where the research was published, psychologists led by the University of Glasgow, in Scotland, reached these findings by asking 80 people (half male and half female, half white European and half East Asian) to observe a number of computer-generated faces and label them as showing either "pain," "orgasm," or "other." The results were used to create a series of better facial animations that more than 100 people then scored and rated.

The findings indicate that people from both cultures associated the same expression—lowered eyebrows, gritted teeth—with the sensation of pain. Why, then, do they appear to associate different expressions with a universal sensation of sexual pleasure?

"It is likely that Westerners and East Asians display different facial expressions in line with the expectations of their culture," the study authors speculated.

"These cultural differences correspond to current theories of ideal effect that propose that Westerners value high arousal-positive states, such as excitement and enthusiasm, which are often associated with wide-open eye and mouth movements, whereas East Asians tend to value low arousal-positive states, which are often associated with closed-mouth smiles.

"That is, Westerners are expected to display positive states as high arousal, e.g., excited, whereas East Asians are expected to display positive states as low arousal, e.g., calm."

The study authors hope that their work will pave the way for more research looking into the role of cultural and perceptual factors in facial expressions, something that is likely to become even easier with the accelerating development of facial recognition technologies.

SCIENTISTS CAN READ RATS' MINDS AND PREDICT WHERE THEY WILL GO NEXT

by Alfredo Carpineti

SCIENTISTS HAVE SUCCESSFULLY DEMONSTRATED MIND-reading. They were only able to read the minds of rats, but, hey, it's a start.

In an area of the brain known as the hippocampus, there are special neurons known as "place cells." These fire up whenever an animal enters a particular place in its environment to construct a "cognitive map" in the brain. In earlier research, this has allowed scientists to figure out where a rat is, based on which neurons are active.

In 2018, researchers also managed to decipher the rat's intentions. As reported in the journal *Neuron*, the team noticed that the activation of a particular place cell could be linked not just to where the rat was, but also to where it was going next.

They placed rats at the center of an eight-way junction, three arms of which contained hidden food, and recorded brain activity in each animal's hippocampus. The team were interested in testing both reference memory, which allows the rat to remember which arms contain food and which don't, and working memory, which allows the rat to remember which arms it has visited and which it still needs to explore.

In reference memory tests, the sequence of place cells firing gave the researchers an idea of the rat's next move. "This gives us an insight into what the animal is thinking about space," senior author Jozsef Csicsvari, from the Institute of Science and Technology Austria, said in a statement.

"We used this concept to understand how rats think during tasks that test their spatial memory. The animal is thinking about a different place than the one it is in. In fact, we can predict which arm the rat will enter next."

The team could even predict more than just where the rat was going. They could also tell when a rat was about to go down the wrong path. "When the rat makes a mistake, it replays a random route," Csicsvari added. "Based on the place cells, we can predict that the rat will make a mistake before it commits it."

97

You Can Spot a Narcissist from This Facial Feature, According to New Study

by Tom Hale

People with "distinctive eyebrows" are more likely to display narcissistic personality traits, suggests research from the University of Toronto, published in 2017 in the *Journal of Personality*. The researchers found that eyebrow thickness and density were correlated with scores on a psychological test for narcissistic personality traits.

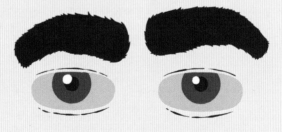

WHY DO YOU LOSE YOUR MEMORY WHEN YOU GET REALLY DRUNK?

by Aliyah Kovner

HAVE YOU EVER HAD TO DEAL WITH SOMEONE WHO IS blackout drunk? Or maybe it's you who's ended up in this state? Either way, the experience may leave you wondering why heavy drinking tends to erase all memory of the experience.

Since we here at IFLScience love answering life's strange questions, let's dive on in.

According to insights from the latest research on the subject, alcohol-induced amnesia is theorized to occur because alcohol, which can cross the blood-brain barrier, interferes with a receptor found on neurons involved in the formation of memories. The inhibitory process—first identified in the early 1990s but still not fully understood—unfolds when ethanol (pure alcohol of the sort found in our drinks) finds its way to the pyramidal neurons within a region of the hippocampus known to be crucial to memory formation. Pyramidal neurons are a type of nerve cell that receives information from other cells before sending one integrated signal onward.

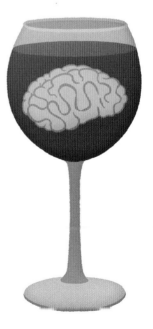

In a 2011 study, published in the *Journal of Neuroscience*, lead investigator Dr. Charles Zorumski and his colleagues used experiments in rats to tease out more details about ethanol's effects. They determined that alcohol messes with a receptor on the pyramidal neurons, causing them to release steroids that inhibit neuron plasticity. And no neural plasticity means no changing synaptic connections, and therefore no memory formation.

"The mechanism isn't straightforward. The alcohol triggers these receptors to behave in seemingly contradictory ways, and that's what actually blocks the neural signals that create memories," Dr. Zorumski said in a statement. "[This finding] may explain why individuals who get highly intoxicated don't remember what they did the night before."

He explained that consumption of copious amounts of alcohol appears to block some receptors while activating others. His group's insights counter the long-held notion that alcohol impairs memory simply by killing brain cells.

"Alcohol isn't damaging the cells in any way that we can detect," Zorumski added. "As a matter of fact, even at the high levels we used here, we don't see any changes in how the brain cells communicate. You still process information. You're not anesthetized. You haven't passed out. But you're not forming new memories."

Past research has shown that acute stress can also impair memory formation, though the process is actually boosted immediately following the threatening event, when corticosteroid hormones involved with the physiological stress response are still high (which makes sense from an evolutionary perspective because animals need to remember and learn from threats they encounter). Some evidence also suggests that long-term stress interferes with memory as well.

Though most of this work has been done in animals, it is predicted to work the same way in humans, meaning that there is a decent probability that if you consume copious amounts of booze, and/or get anxious, then your brain's ability to store new information will be compromised.

This Type of Man Gives the Best Orgasms
by Dami Olonisakin

In 2017, researchers surveyed 103 single women, aged 20 to 69, about what kind of man produced the best orgasms. According to their results, it wasn't all a matter of good looks. The best experiences were with men having positive character traits—being more humorous, creative, warm, faithful, and better smelling than partners who induced low-orgasm rates.

These Personality Traits Could Dictate How Often Men Have Sex, Study Claims
by Jonathan O'Callaghan

A study by Queensland University of Technology in Australia published in the journal *Personality and Individual Differences*, surveyed 3,000 heterosexual males about their personality traits and how often they had sex. It found that men who were more extroverted, conscientious, emotionally stable, but less agreeable, tended to have sex more often.

$$h/(2M_nE)^{1/2} \quad K = P^2 \frac{\sin\beta}{2m} \quad V_2 \quad m_1 \quad P = UI \quad \Phi = NBS$$

$$E_0 \sin(kx - \omega t) \quad R_m = \frac{C}{T} \quad R = \frac{U}{I} \quad \frac{\Delta\varphi}{2\pi} = \frac{\Delta x}{\lambda} = \frac{x_2 - x_1}{\lambda}$$

$$\frac{\Delta}{f_1} \cdot \frac{\ell}{f_2'} = z_1 z_2 < 0 \quad \Delta t = \frac{\Delta t'}{\sqrt{1 - \frac{V^2}{c^2}}} \quad \gamma = \frac{tg\tau'}{tg\tau} \quad \frac{d}{f} \quad V = V_1(1+$$

$$z = z_{ob} \cdot \mu_{ok} = \frac{\Delta}{f_1} \cdot \frac{d}{f_2} \quad P = \frac{E}{c} = \frac{hf}{c\Delta E} = \frac{h}{\lambda} \quad f_0 = \frac{1}{2\pi\sqrt{CL}} \quad M_e = \sigma T \quad m = \Lambda$$

$$\sqrt{\frac{3kTN_A}{M_m}} = \sqrt{\frac{3R_mT}{M_R \cdot 10^{-3}}} \quad P = \frac{\vec{F}}{\Delta S} = \frac{m\Delta \vec{V}}{\Delta S \Delta t} \quad U_{ef} = \frac{U_m}{\sqrt{2}} \quad h = \frac{1}{2}g$$

$$\sqrt{\left[\frac{1}{R^2} + \left(\frac{1}{X_C} - \frac{1}{X_L}\right)^2\right]} \quad \Delta\Psi = \frac{2\pi\Delta x}{\lambda} = \frac{2\pi d\sin\vartheta}{\lambda} = \frac{2\pi d}{x}$$

$$\oiint \vec{D}d\vec{S} = Q^* \quad X_L = \frac{U_m}{I_m} = \omega L$$

$$W = F \cdot s \cdot \cos\alpha \quad W_2 = U_e I t \quad V = \frac{nh}{2\pi r m_e} \quad \phi_e = \frac{L}{4\pi r^2}$$

$$E_k = \frac{h^2}{8mL^2} \quad \sigma = \frac{Q}{S} \quad \vec{B} = \mu \frac{NI}{\ell} \quad U = \frac{W_{AB}}{Q}$$

$$r^2 = M_z \frac{4\pi^2 r}{T^2} \quad \beta = \frac{\Delta I_C}{\Delta I_B} \quad tg\vartheta_B = \frac{n_2}{n_1} = n_{21}$$

$$F_x = \frac{1}{2}C_x \rho S \vec{v}^2 \quad E = mc^2$$

$$1 pc = \frac{1 AU}{r} \quad E = \frac{E_c}{a}\int_{-a/2}^{+a/2} \sin(\omega t + \phi)dy \quad \oint_{c(s)} \vec{H}d\vec{\ell} = \iint_S$$

$$n\omega(t-\tau) = U_m \sin 2\pi\left(\frac{t}{T} - \frac{x}{\lambda}\right) \quad E_k = \frac{1}{2}mv^2 \quad \lambda = \frac{\ln_2}{T} \quad F_g$$

$$\iint_S \frac{\partial \vec{B}}{\partial t}\cdot d\vec{S} \quad S = \frac{1}{A}\frac{d\omega}{dt} \quad \vec{\Psi} = \iint_{S_2}\vec{D}d\vec{S} = AD \quad \left(\frac{E_t}{E_0}\right.$$

$$\mu \iint_S \vec{J}d\vec{S} \quad \nabla\times\left(-\frac{\partial\vec{B}}{\partial t}\right) = -\frac{\partial}{\partial t}(rot\vec{B}) = -\mu_0\frac{\partial}{\partial t}\left(\frac{\partial\vec{B}}{\partial t}\right)$$

$$F_e = k\frac{Q_1 Q_2}{r^2} \quad \frac{n_1}{X} + \frac{n_2}{X'} = \frac{n_2 - n_1}{R} \quad \vec{S} =$$

$$n(kx - \omega t)$$

CHAPTER SEVEN

PHYSICS AND CHEMISTRY

ERNEST RUTHERFORD, THE NEW ZEALAND-BORN British physicist who became known as the "father of nuclear physics" after correctly sussing out the structure of the atom in 1911, once commented that "Physics is the only real science. The rest are just stamp collecting."

The line is frequently trotted out by smug physicists hoping to taunt their colleagues in other scientific disciplines.

Rutherford's point was that physics is the most fundamental of the sciences, and he believed that made it the only science with any true explanatory power. The rest just amounted to classification schemes, and some other laws that were ultimately derived from physics.

Physics, in essence, is a set of mathematical laws describing how space, time, energy, and elementary particles of matter all behave. And, if you believe Ernest, then from that everything else follows. Physics tells you how atomic nuclei combine with electrons to form atoms. And it says how those atoms interact with one another when they come together.

That's pretty much what the science of chemistry is all about. When, for example, hydrogen gas burns, that's a chemical reaction where atoms of hydrogen and oxygen combine, sharing some of their electrons and releasing energy in the process. Yes, this is chemistry but it all comes from the laws of quantum physics, which govern how atoms and electrons go about their daily business. And you can take the chain of reasoning further. The science of biology is really just a subset of chemistry, dealing with the particular group of chemicals, and the structures they form, that maintain and propagate life.

Taking Rutherford at his word, however, things do start to get rather circular. Life (biology) comes from physics. Some life is human life, and some humans are physicists . . . So, eventually, we reach the inescapable conclusion that physics explains physicists, who explain physics. And what about free will, if that's even a thing? If everything is predetermined by physics, then how do I at least appear to decide for myself when I'm going to watch TV, eat dinner, or go to the toilet?

Anyway. For now, let's ignore such philosophical quandaries and concentrate instead on the cool bits of physics, and its close cousin chemistry—those two old favorites from high school science classes, where you'd get to see demonstrations involving things being variously squashed, snapped, burned, blown up, electrocuted, and irradiated.

We should add that there's often been a kind of cozy, homespun disregard for health and safety in physical science research over the years, as wild-haired yet lovable professors tested their crazy theories with little concern for their own safety, or indeed that of others. Nothing exemplifies this better than when Enrico Fermi built the world's first nuclear reactor, in an unused squash court at the University of Chicago. The reactor's safety system consisted of a man with an axe—whose job it was, in the event of a problem, to cut the rope suspending an emergency control rod over the reactor core.

Among the scintillating physics and chemistry stories that we detail here, find out how scientists have detected antimatter inside a thunderstorm. Particles of antimatter are like the mirror image of ordinary matter particles, having their key properties, like electric charge, reversed. When matter and antimatter meet, the result is a powerful release of energy—so you might reasonably have expected the stuff to only crop up inside particle accelerators, or on the event horizon of a black hole. These terrestrial antimatter particles are thought to have been created by the huge energies generated by electric fields within the storm.

We meet the gamers who helped to prove Albert Einstein wrong, participating in a massive worldwide experiment into the nature of quantum theory. Despite his pioneering insights into relativity, Einstein always hated quantum mechanics—developing his own theories to explain some of the weirder goings-on in the particle world. Now, over 100,000 people have taken part in a cleverly engineered online game, the results of which were, basically, Quantum Mechanics 1–Einstein 0.

And we look at the latest designs for the successor to the Large Hadron Collider (LHC), a new particle accelerator, called the Future Circular Collider, that's due to be switched on around 2040–2050. It will be four times the size of the LHC and will generate particle collision energies that are 10 times higher.

We've also got some fairly awesome chemistry stories for you. Find out how wannabe Willy Wonkas have made a fourth type of chocolate, called "ruby" (in addition to dark, milk, and white)—and, yes, you can buy it on Amazon. Find out how to cook up gold that's golder than gold itself. And read about the researchers who claim to have invented a working *Star Wars* "moisture vaporator"—a device that can extract moisture from the air, providing a potentially lifesaving source of water in arid climates.

A word of warning though: Undertaking chemistry on a DIY basis is often asking for trouble, as was demonstrated in 2018 by the well-meaning lady in the town of Nailsea, England, who attempted to unblock a toilet using two bottles of sulfuric acid and several quarts (liters) of bleach, and inadvertently created a toxic cloud of chlorine gas (very nasty, banned by the Geneva Protocol) that proceeded to envelope her street. We know what you're thinking—where do I get sulfuric acid? Apparently, vinegar works quite well, too.

PHYSICS AND CHEMISTRY

101
CHINA JUST SET A NEW NUCLEAR FUSION RECORD BY REACHING TEMPERATURES OF 180 MILLION DEGREES
by Jonathan O'Callaghan

CHINA SAYS IT HAS PERFORMED A TEST OF A NUCLEAR fusion machine it is developing, reaching temperatures seven times hotter than the center of the Sun.

The test took place at the Experimental Advanced Superconducting Tokamak (EAST) at the Hefei Institutes of Physical Science of the Chinese Academy of Sciences (CASHIPS). The machine reached temperatures of more than 180 million °F (100 million °C), the temperature at which nuclear fusion takes place. EAST is a tokamak reactor, which is shaped like a doughnut and uses large electric currents to twist the plasma inside, confining it using magnetic fields.

In the past few years, a number of experimental fusion reactors have successfully sustained a plasma for a minute or so. China's test is particularly significant, though, for the temperature it reached. The interior of our Sun reaches temperatures of "just" 27 million °F (15 million °C). But to kick-start nuclear fusion in a reactor on Earth, temperatures about seven times higher are required.

"If we can achieve that, the payoff would be massive," *ScienceAlert* noted. "Unlike nuclear fission—where surplus energy comes from the decay of large atoms into smaller elements—nuclear fusion doesn't result in anywhere near as much radioactive waste. In fact, the end result of squeezing together isotopes of hydrogen is mostly helium."

China had previously set a world record of sustaining a plasma for 101.2 seconds last year, and has now turned its attention to raising the temperature inside the machine. The ultimate goal will be to sustain this plasma indefinitely, providing a clean and practically endless source of power.

China is part of an international collaboration known as ITER (International Thermonuclear Experimental Reactor), along with 34 other countries, to develop an operational fusion power plant. The experiments at EAST go some way toward making that dream a reality.

"EAST, a device independently designed and developed by Chinese scientists to harness the energy of nuclear fusion, is taking a step closer to maintaining a more stable fusion reaction as long as possible and at an even higher temperature," the Chinese state-owned CGTN news network said.

PHYSICS AND CHEMISTRY

THESE SCIENTISTS SAY THEY'VE INVENTED SOMETHING THAT CAN CREATE WATER OUT OF DESERT AIR

by Katie Spalding

OUR PLANET IS NOTHING IF NOT IRONIC. EARTH IS covered in water—millions of trillions of gallons (liters) of the stuff—and yet only 2 percent of this is drinkable. Of that, 99.5 percent is frozen or buried below the ground. And of what's left—well, human-made climate change is taking care of that.

One piece of good news, however, is that water we can drink isn't just confined to places like lakes, rivers, and raindrops. There's almost 13 trillion tons of delicious H_2O hidden in plain sight all around us—in the air. We just have to extract it.

There are a few ways to do this, but most are either too inefficient or prohibitively expensive. However, researchers in Saudi Arabia say they have a solution: a simple device that can harvest and store its own weight in water, and release it when warmed by sunlight.

The key to the prototype is a cheap, stable, eco-friendly, and non-toxic chemical salt known as calcium chloride. This salt is so good at absorbing water that it will literally dissolve if left in fresh air—a property known to chemists as deliquescence.

But calcium chloride turns liquid after absorbing water, which is a problem. To combat this, the team developed a way of storing the calcium chloride as a hydrogel—a special type of polymer that can hold vast amounts of water while remaining solid. And with the addition of some tiny carbon nanotubes to let the water escape, team members were able to use a light source to reclaim almost 100 percent of the water captured by the gel.

In a paper, published in *Environmental Science and Technology* in 2018, the team described the results of their small, "easy-to-assemble-at-household" prototype. A device incorporating 1.2 ounces (35 grams) of hydrogel absorbed 1.3 ounces, or about 7.5 teaspoons (37 grams), of water when left overnight in air with a relative humidity of 60 percent. And nearly all of this water was later released and collected by the device after 2 ½ hours' exposure to sunlight.

The researchers say this could be scaled up to provide an adult's minimum water requirement for a day—6.6 pounds, or 4 ¼ cups (3 kg)—with a daily running cost of just half a cent.

Along with its low cost and high water yield, the device has the advantage of working well even in relatively low humidity—perfect for arid or drought-stricken regions. It also needs no electricity, meaning it can be used even in the remotest parts of the world.

"Water scarcity is one of the most challenging issues that threaten the lives of mankind," the paper reports. "This technology provides a promising solution for clean water production in arid and land-locked remote regions."

World War II Bombing Raids Were Felt Even at the Edge of Space

by Alfredo Carpineti

Bomb blasts during World War II reached the upper atmosphere. That's according to a team who have matched 1940s observations of Earth's ionosphere—the layer of atmosphere between 37 and 620 miles (60–1,000 km) up—to the known times of 152 Allied air raids.

IFLSCIENCE

AMATEUR SCIENTISTS JUST PROVED EINSTEIN WRONG

by Alfredo Carpineti

PROVING ALBERT EINSTEIN WRONG IS SOMETHING THAT only a small number of scientists can claim to have done. But now, more than 100,000 gamers can join that exclusive club and enjoy the smugness that comes with it.

In 2016, scientists from around the world, led by the Institute of Photonic Sciences in Barcelona (ICFO), asked people to play a simple game online, and the results were used to disprove one of Einstein's claims about quantum mechanics, the branch of physics governing atoms and subatomic particles. The results were published in 2018 in the journal *Nature*.

One of the things Einstein truly disliked about quantum mechanics is how the experimenter plays a role in the results obtained from an experiment. He believed the universe to be independent of our actions and quantum mechanics to be governed by the "principle of local realism."

This principle tells us that there should be hidden variables in the theory that can explain puzzling effects like entanglement—where two particles are seemingly able to influence one another, regardless of the separation between them—which Einstein dismissed as "spooky action at a distance."

However, quantum mechanics seems to work fine without local realism and scientists have proved this using the so-called Bell test experiment. Here, two entangled particles are sent to different locations and their properties are measured. The measurements of one particle influence the other in this kind of experiment, which, according to Einstein, suggests there's some sort of faster-than-light communication going on. But in quantum mechanics, all is fine—entangled particles are a single system that shouldn't be considered independently, no matter how far they are separated. Time and again, the results of the Bell test experiment have supported quantum mechanics over local realism.

But one limitation of the test has remained—what's called the "freedom-of-choice loophole," where it's been argued that the choice of which specific particle properties to measure might influence the outcome of the experiment. To exclude this possibility, the team needed to randomly determine which measurements to make—using random number generators that were completely independent of the system. And that's where the general public came in.

The project, ambitiously named the BIG Bell Test, recruited over 100,000 people to play simple games that created long strings of zeros and ones. These "bits" (short for "binary digits") were then routed to 12 labs across the world where they were used as random numbers. The live results contradicted local realism with 99.7 percent confidence.

"The BIG Bell Test was an incredibly challenging and ambitious project," Carlos Abellán, a researcher at ICFO and instigator of the project, said in an emailed statement. "It sounded impossibly difficult on day zero, but became a reality through the efforts of dozens of passionate scientists, science communicators, journalists, and media, and especially the tens of thousands of people that contributed to the experiment during November 30, 2016."

IFLSCIENCE

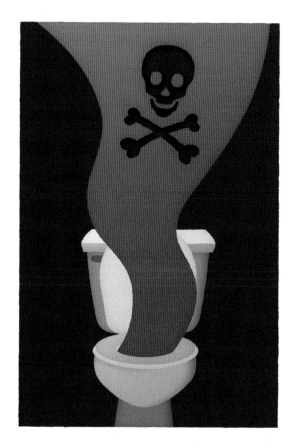

RESIDENTS OF UK TOWN FORCED TO EVACUATE AFTER CLEANING ACCIDENT GOES VERY WRONG

by Rosie McCall

RESIDENTS OF NAILSEA, A BRITISH TOWN NEAR BRISTOL, had a somewhat unusual end to 2018 when they were forced to evacuate their homes after a noxious gas cloud erupted on their street.

The good news is that this was not a preplanned chemical attack, targeting the people of Nailsea. Rather, it turned out to be a simple cleaning accident by a woman attempting to unblock a toilet.

Dominique Heath, a mom of three, had the unpleasant task of unclogging a toilet on the day after Boxing Day when one of her children blocked it with a toy or too much paper. In the afternoon, she poured two bottles of a toilet unblocker down the pan and left it to do its job. But at 8 p.m., the toilet was still clogged. And so she added a 0.8-gallon (3-liter) tub of bleach to the mix. For the record, that is a lot of cleaning solution.

Unfortunately for Heath, the chemicals in the toilet unblocker (principally, sulfuric acid) and the bleach (sodium hypochlorite) reacted to form a putrid cloud of chlorine gas, a poisonous substance so toxic it is banned by the Geneva Convention and its use is considered a war crime. Germany deployed it to devastating effect in 1915 at the Second Battle of Ypres during World War I.

In small doses, it can cause skin and eye irritation but too much of it can lead to chemical burns and breathing problems. In the very worst cases, it can result in non-cardiogenic pulmonary edema (buildup of fluid in the lungs), which can be lethal.

Undoubtedly a little concerned about the foul-smelling gas, Heath shut the bathroom door and opened up all the windows in the house, she told *Bristol Live*.

"I have never experienced fumes like it," she said. "My throat and eyes feel burned."

The cloud continued to spread so she turned to a neighbor for help and called the fire department, who immediately told her to evacuate. They then sent fire crews from three separate stations in the Bristol and North Somerset area to attend to the chemical accident and cordon off the end of the street until it was safe to return.

Since the whole affair came to a close, Heath has shared her story on Facebook, saying she wants to use her experience as a warning to anyone else who might be tempted to mix large quantities of potentially lethal chemicals.

"It was really serious," she said, *Bristol Live* reports. "We are all okay, but it was the dumbest thing I have ever done—please don't do this!"

Fortunately, it appears no one was hurt.

New Form of Lab-Made Gold Is Better and Golder Than Nature's Pathetic Version

by Robin Andrews

Gold is known for being ultraresistant to interactions with other chemicals. In 2018 researchers made a kind of gold in the lab that's even more resistant than usual. They cooked it up by heating gold chloride in a furnace at 428°F (220°C) for half an hour.

PHYSICS AND CHEMISTRY

WHY DOES THE SOUND OF YOUR OWN RECORDED VOICE BOTHER YOU SO MUCH?

by Tom Hale

THERE'S NOT A PERSON ALIVE WHO LIKES TO LISTEN TO A recording of his or her own voice. Even if you sound like honey poured over thunder, the voice you hear played back on a speaker certainly doesn't sound like the voice you've heard coming out of your mouth for all these years.

This strange phenomenon isn't you being insecure about your vocal inadequacies. No, there's a logical reason why your voice sounds so different—and god-awful—on recordings.

We hear sounds through vibrations being picked up by our eardrum. The vibrations are sent to three bones in the middle ear and then finally to the cochlea, a snail-shaped organ that turns the vibrations into nerve signals.

We perceive external sounds, like a beeping car or a radio, via sound waves passing through the air into our ear canals, then into our inner ear, and on to the cochlea. When our voice is played back to ourselves on a speaker, we are hearing these air-conducted vibrations.

It's a bit different, however, if the sound is coming from our own vocal cords. A lot of what we hear when we speak is perceived in the same way as external noise, but we also pick up on vibrations that have come through our jawbone and skull. This is known as inertial bone conduction, an effect you can demonstrate if you bang a tuning fork and place the handle against your skull.

This also alters the quality of the sound you hear. Bone conduction tends to "bring out" the lower-frequency vibrations, making your voice sound deeper and less squeaky than it actually is. In all likelihood, the fact that you don't like the sound of it is simply because you are not used to it.

You can also try out the opposite effect for yourself by sticking your fingers in your ears, so you filter out air-conducted sound altogether and hear only the bone-conducted vibrations. You'll notice your voice sounds a lot deeper than normal.

Of course, the depressing reality is that the awful noise you hear when you play back a recording of your voice is actually how your voice sounds to the 7.6 billion other humans on Earth.

Sorry about that, folks.

SCIENTISTS SAY THEY'VE CREATED A STRANGE NEW STATE OF MATTER THAT DOESN'T PLAY BY THE RULES

by Jonathan O'Callaghan

Scientists have described an odd new state of matter that doesn't seem to conform to normal rules.

As reported by *New Scientist* in October 2018, the state is made from "liquid light" and is classed as being between a solid and a superfluid, a liquid that flows with zero friction.

The truly weird thing about the new substance is that it is completely rigid, meaning it can't be pushed or stirred. It's made of "polaritons," which are hybrid particles composed "of a photon strongly coupled to an electric dipole," explained *Nature*.

The team made the fluid by trapping light in another material, using a laser to replace any light photons that leaked out of the experiment. The research was published in *Nature Communications*, and conducted by scientists at University College London and the University of St Andrews.

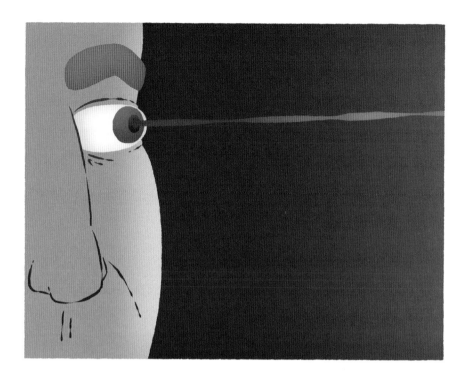

PEOPLE SECRETLY BELIEVE THAT THE EYES SEND OUT FORCE-CARRYING BEAMS

by Stephen Luntz

OLD IDEAS DIE HARD, INCLUDING THE BELIEF THAT THE eyes send out invisible beams that can affect what we are looking at.

When Professor Michael Graziano of Princeton University asked 724 participants in a study if they believed people could exert forces with their eyes, only 5 percent said yes. After all, philosophers have reasoned for 2,500 years that the eyes must work because of light entering them, not leaving. However, Graziano reports in *Proceedings of the National Academy of Sciences* that when he tested a subgroup's subconscious attitudes, the results were quite different.

Graziano showed 157 subjects images of paper tubes of different heights and diameters, and asked them to guess how far they could be tilted without falling over. The images included a photograph of a man referred to as Kevin staring at the tube, but sometimes he was blindfolded.

When Kevin's eyes were open, participants thought the tube could be tilted more strongly toward him than away. For images where he was blindfolded, or when told the tube was made of very heavy material, this bias disappeared. The responses might have been correct if Kevin were producing invisible beams from his eyes, which could support the tube when it tilted toward him, and help push it over when the tilting went the other way.

The difference in responses was small—just 0.64° on average—but highly statistically significant. Graziano and co-authors calculated that the estimates would be right if Kevin's eyes exerted a force of less than one-hundredth of a Newton, "similar in magnitude to a barely detectable breeze." Even discounting the seven subjects who openly believed that the eyes exert a force still didn't change the results.

The authors note that there is an extraordinary persistence across cultures of "the belief the eyes emit an invisible energy," something that's also assumed by almost all children. Scientific demonstrations that this isn't true seem to be fighting against deeply embedded presumptions.

Graziano is known for his original approaches, including performing ventriloquism in lectures using an orangutan puppet, also named Kevin. He uses the way our brains attribute consciousness to puppets as an example of the way humans as social creatures project models of others' thoughts.

We're sure we'd have paid more attention in our university classes if they'd been taught anything like that.

IFLSCIENCE

CHINESE AND RUSSIAN SCIENTISTS HAVE BEEN HEATING HUGE PORTIONS OF THE ATMOSPHERE (ON PURPOSE)

by Katie Spalding

ABOUT NINE HOURS EAST OF MOSCOW, WHERE THE SURA River meets the Volga, there's an ex-Soviet laboratory pumping high-frequency radio waves into the upper atmosphere.

As the waves reach the ionosphere, they disrupt it in mysterious and unexplained ways, forming artificial plasma ducts and heating the ions and electrons that make up this atmospheric region.

It sounds like the opening to an apocalyptic blockbuster, but experiments like these aren't unusual. Northeast of Anchorage, Alaska, the US has its own facility, the High-frequency Active Auroral Research Program, or HAARP. These days, the research station belongs to the University of Alaska Fairbanks, but for nearly a quarter of a century it was the property of the US Air Force and Navy, bombarding the atmosphere with radio waves in the search for new or improved military applications.

This is because the ionosphere plays an important role in communication. Ionized particles reflect radio waves sent from Earth and can disrupt signals from satellites—control the ionosphere, and you can potentially cut off your enemies' access to information. It's no surprise, therefore, that it's been a priority for military forces—as well as the source of quite a few conspiracy theories—around the world.

So the news that Russian and Chinese scientists have been collaborating on a project to do just that has understandably turned some heads. As reported by the *South China Morning Post*, five experiments were carried out in the Russian skies over a period of 11 days in June 2018, sending high-frequency waves into the atmosphere with enough power to light a small city.

"We are not playing God," one researcher promised the *Post*. "We are not the only country teaming up with the Russians. Other countries have done similar things."

In a paper published in December 2018 in the journal *Earth and Planetary Physics*, researchers from the Institute of Earthquake Forecasting in Beijing and the Russian Radiophysical Research Institute described how the in-orbit China Seismo-Electromagnetic Satellite (CSES) was directed to monitor a series of high-frequency radio waves sent into the ionosphere from the Russian Sura research lab.

The paper describes one experiment that increased the temperature in the upper atmosphere by more than 180°F (100°C), while the *Post* reports that another caused a physical disturbance as large as 49,000 square miles (126,000 km^2)—about the size of Mississippi.

Northern Lights are formed when the atmosphere is bombarded by charged particles from the Sun.

Despite the alarming headlines, experts say experiments like these are subject to strict guidelines. "Such studies must strictly follow ethical guidelines," Gong Shuhong, a military communication technology researcher, explained to the *Post*. "Whatever they do, it must not cause harm to the people living on this planet."

And although the unusual level of cooperation between the two governments may have some worried about malicious agents jamming communication signals, the researchers stress that the experiments were carried out in the name of science alone.

"We are just doing pure scientific research," study author Wang Yalu told the *Post*. "If there is anything else involved, I am not informed about this."

So there you have it. Do not adjust your set.

Scientists Have Invented a Fourth Type of Chocolate
by Robin Andrews

Pink-colored "ruby chocolate" is manufactured from the ruby cocoa bean. It is the first new variety since white chocolate was introduced to the world 80 years ago, and brings the number of varieties to four. Ruby chocolate was launched in 2017 by Barry Callebaut, a Zürich-based company, after 13 years of development. And yes, it's available on Amazon.

PHYSICS AND CHEMISTRY

STUDY CLAIMS THAT THE GREAT PYRAMID OF GIZA COULD FOCUS SOME ELECTROMAGNETIC WAVES

by Alfredo Carpineti

RESEARCHERS HAVE DISCOVERED SOMETHING PECULIAR about the Great Pyramid of Giza in Egypt—the shape is ideal for focusing certain electromagnetic waves. And while this discovery is unlikely to provide any insights into ancient Egyptian burial practices, it might lead to the creation of better nanotech.

The study, published in the *Journal of Applied Physics*, looked at how the shape of the pyramid would interact with resonant radio waves. Resonance is an important physical phenomenon, where small wave oscillations are able to pile up on top of one another to create large oscillations. Given the pyramid's size and shape, the resonant radio waves are expected to have wavelengths between 650 and 1,970 feet (200–600 meters).

The team found that external waves of the resonant wavelength striking the pyramid were concentrated by its shape, focused either into the pyramid's central chambers or into the region below its base, depending on the specific model for its structure that they used.

Just to be clear, this study is about theoretical calculations and shouldn't be taken as an explanation for some profound mystery surrounding Egypt's tombs. The pyramid most definitely can't receive or send alien messages because our atmosphere is not transparent to signals at the resonant wavelengths.

"Egyptian pyramids have always attracted great attention. We as scientists were interested in them as well, so we decided to look at the Great Pyramid as a particle dissipating radio waves resonantly," senior author Dr. Andrey Evlyukhin said in a statement.

"Due to the lack of information about the physical properties of the pyramid, we had to use some assumptions. For example, we assumed that there are no unknown cavities inside, and the building material with the properties of an ordinary limestone is evenly distributed in and out of the pyramid. With these assumptions made, we obtained interesting results that can find important practical applications."

The team plans to exploit these properties of pyramid-shaped objects in nanostructures. They hope to use the focusing ability in sensors and solar cells—hopefully using materials better suited for these functions than the limestone used to build the Great Pyramid of Giza.

Why Do You Get More Static Electric Shocks When It's Cold?

by Tom Hale

Static electricity is the buildup of electric charges on a material. When air is moist, its conductivity increases, making it easier for static charges to dissipate away. Cold air, however, holds less moisture, so its conductivity is lower—and this makes it more likely that charge will accumulate to the point where you get a shock.

PHYSICS AND CHEMISTRY

HURRICANE PATRICIA'S LIGHTNING FIRED A BEAM OF ANTIMATTER DOWN TO EARTH

by Robin Andrews

A NEW PAPER, PUBLISHED IN THE *JOURNAL OF GEOPHYSical Research: Atmospheres*, has quite the title. Obviously you shouldn't judge a book by its cover, but we'd argue this one warrants close attention: "A terrestrial gamma-ray flash inside the eyewall of Hurricane Patricia."

In other words, this means that one of the windiest hurricanes on record produced a fair bit of lightning, and at least one of those flashes was energetic enough to produce a beam of antimatter, which shot down to Earth. If you don't think that's cool, then that's fine, but you'd be wrong.

Back in 2015, when Hurricane Patricia was wreaking havoc on Mexico's west coast, records were being set. The intensity of this cyclonic behemoth was unparalleled, and scientists wanted to get a closer look. Thankfully, the US National Oceanic and Atmospheric Administration (NOAA) has some specially designed aircraft that can fly into the hearts of hurricanes unharmed.

As one of NOAA's Hurricane Hunter planes flew directly into Patricia's peak paroxysmal rage, it headed for the eyewall, a circumference of colossal thunderstorms, high winds, and extreme weather.

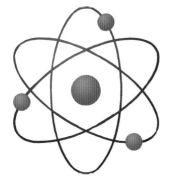

An instrument aboard the craft named ADELE—the Airborne Detector for Energetic Lightning Emissions—designed by engineers at the University of California, Santa Cruz, picked up 184 counts of ionizing "gamma" radiation in the blink of an eye, following a lightning flash. Based on the associated radio signal, and comparing the gamma-ray energy spectrum to simulations, the team concluded that ADELE had happened upon a beam of positrons.

Positrons are the antimatter equivalents of electrons; they have the same mass, but an equal yet opposite charge. So what's the science behind this spectacular occurrence of lightning-generated antimatter?

The average lightning bolt involves the transfer of a billion or so joules of energy, equivalent to detonating around a quarter of a ton of TNT. This enormous energy release can accelerate charged particles to terrific speeds, and these high-energy particles then emit gamma rays as they smash into atoms in the atmosphere. The collisions rip further electrons from the atoms, which themselves collide, generating more gamma rays in a cascade effect—producing a terrestrial gamma-ray flash (TGF). The gamma rays soon decay, in a process called "pair production," back into electrons and their antimatter partners, positrons. This creates a beam of electrons that shoots upward into space, just as the oppositely charged positrons move downward, their negative charge sending them in the opposite direction in the thunderstorm's electrical field.

Fortunately, the event detected by NOAA in 2015 has been found to match the models perfectly: ADELE picked up a downward-blasted beam of positrons. Detecting this TGF-linked antimatter beam wasn't a surprise; it is, however, the first time it's actually been observed, which is clearly marvelous. In case you're wondering: No, you'd have to be right near the source of this beam for it to be of any danger.

This certainly won't be the last time an antimatter beam like this will be detected. You might not even need to brave flying into hurricanes; the team's study explained that "this reverse gamma-ray beam penetrates to low enough altitudes to allow ground-based detection of typical upward TGFs from mountain observatories." Not quite as cool as flying into a hurricane, though.

Quantum Uncertainty Suggests That an Object Can Be Multiple Temperatures at Once

by Alfredo Carpineti

The "uncertainty principle" of quantum mechanics says it's impossible to pin down the position and momentum of a subatomic particle of matter—it can be thought of as in several places at once. A study in 2018 in *Nature Communications* extended this concept to temperature, so particles in effect have several temperatures at once.

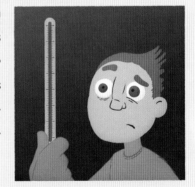

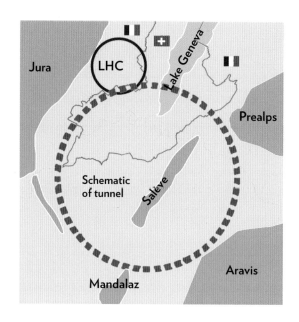

CERN ANNOUNCES CONCEPT DESIGN FOR ITS 62-MILE (100-KM) FUTURE COLLIDER

by Alfredo Carpineti

CERN'S LARGE HADRON COLLIDER (LHC) HAS EXPANDED our understanding of fundamental physics significantly, thanks to the discovery of the Higgs boson. But while the LHC still has much to give, physicists have already started thinking about the next big thing, and it's called the Future Circular Collider (FCC).

The conceptual design report for the ambitious project has been released and it combines the work of 1,300 collaborators from 150 universities, research institutes, and industrial partners to deliver a series of concepts for what the FCC might look like. The accelerator is envisioned to be 62 miles (100 km) long, almost four times the size of the present LHC.

The team estimates a 10-fold increase in the energy of particle collisions. This would allow us to study the interaction between the Higgs boson and other particles, helping us to understand the behavior of matter in the early universe and even look for new massive particles at higher energies. The FCC will put the standard model of particle physics to the most exacting tests yet.

"The FCC conceptual design report is a remarkable accomplishment. It shows the tremendous potential of the FCC to improve our knowledge of fundamental physics and to advance many technologies with a broad impact on society," CERN Director-General Fabiola Gianotti said in a statement. "While presenting new, daunting challenges, the FCC would greatly benefit from CERN's expertise, accelerator complex, and infrastructures, which have been developed over more than half a century."

The current plan will be discussed in the context of the European Strategy for Particle Physics, alongside other proposals such as the Compact Linear Collider (CLIC). The proposal calls for the construction of the tunnel and a positron-electron collider, potentially beginning by 2040–2050. This collider will serve the particle community for 15 to 20 years, after which a proton collider will be installed in the tunnel.

"Proton colliders have been the tool of choice for generations to venture new physics at the smallest scale," said Eckhard Elsen, CERN director for research and computing. "A large proton collider would present a leap forward in this exploration and decisively extend the physics program beyond results provided by the LHC and a possible electron-positron collider."

The full proposal is contained in four volumes and took five years to complete. It has received the strong support of the European Commission.

IFLSCIENCE

Why 117 things?

No reason really. We just liked the number. Although it may interest you to know that in November 2016, the International Union of Pure and Applied Chemistry (IUPAC) assigned element 117 of the periodic table—that is, a chemical element with 117 protons in its nucleus—the name tennessine (pronounced "ten-ess-een"). That's after the state of Tennessee—home of Elvis, Jack Daniels, and now the lab where this little element was discovered.

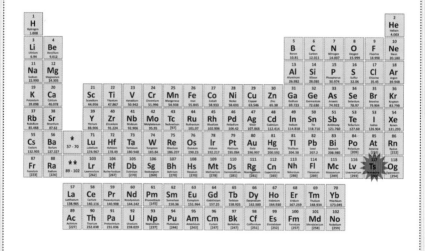

APPENDIX

The articles featured in this book have been adapted from original IFLScience content. Some have been shortened from their original form, but can be found in their entirety at the web addresses listed below.

1) **"Suicide Machine" That Lets You Experience Death Is Now Available for the Public to Try**
https://www.iflscience.com/technology/experience-in-virtual-reality-what-it-feels-like-to-die-in-worlds-first-suicide-machine/
by Madison Dapcevich • Originally published April 16, 2018

2) **Pornhub's Data Shows Something Hilarious Happened during the Royal Wedding**
https://www.iflscience.com/editors-blog/pornhub-data-reveals-the-world-chose-a-bizarre-way-to-honor-the-royal-wedding/
by James Felton • Originally published May 29, 2018

3) **Scientists Have Created AI Inspired by HAL 9000 from *2001: A Space Odyssey***
https://www.iflscience.com/technology/scientists-have-created-ai-inspired-by-hal-9000-from-2001-a-space-odyssey/
by Tom Hale • Originally published November 23, 2018

4) **The US Has Lost Five Nuclear Weapons. So Where the Hell Are They?**
https://www.iflscience.com/technology/the-us-has-lost-six-nuclear-weapons-so-where-the-hell-are-they/
by Tom Hale • Originally published May 4, 2018

5) **People Who Use Someone Else's Netflix Password, We've Got Bad News**
https://www.iflscience.com/technology/people-who-use-someone-elses-netflix-password-weve-got-bad-news/
by Tom Hale • Originally published January 10, 2019

6) **Scientists Say They Can Turn Human Poop into Fuel**
https://www.iflscience.com/technology/scientists-say-they-can-turn-human-poop-into-fuel/
by Jonathan O'Callaghan • Originally published November 20, 2018

7) **Facebook Has a "Secret File" on You. Here's How You Can View It**
https://www.iflscience.com/technology/heres-how-to-find-out-everything-facebook-knows-about-you/
by Jonathan O'Callaghan • Originally published March 2, 2018

8) **Self-Driving Tesla Mows Down and "Kills" AI Robot at CES Tech Show**
https://www.iflscience.com/technology/selfdriving-tesla-mows-down-and-kills-ai-robot-at-ces-tech-show-/
by Rosie McCall • Originally published January 8, 2019

APPENDIX

9) **Thousands of Missing Children in India Identified through Facial Recognition Pilot Experiment**
https://www.iflscience.com/technology/thousands-of-missing-children-in-india-identified-in-facial-recognition-pilot-experiment/
by Aliyah Kovner • Originally published April 24, 2018

10) **Scientists Have Created a *Star Trek*–Like Plane That Flies Using "Ion Thrusters" and No Fuel**
https://www.iflscience.com/technology/scientists-have-created-a-star-treklike-plane-that-flies-using-ion-thrusters-and-no-fuel/
by Jonathan O'Callaghan • Originally published November 21, 2018

11) **Your Dream of Being Able to Breathe Underwater May Soon Be a Reality**
https://www.iflscience.com/technology/your-dream-of-being-able-to-breath-underwater-may-soon-be-a-reality/
by Katy Evans • Originally published August 17, 2018

12) **IBM Has Built the "World's Smallest Computer" That Can Be Put "Anywhere and Everywhere"**
https://www.iflscience.com/technology/ibm-unveils-worlds-smallest-computer-can-be-put-anywhere-and-everywhere/
by Robin Andrews • Originally published March 20, 2018

13) **This Is What Happens When a Drone Slams into a Plane's Wing at High Speed**
https://www.iflscience.com/technology/this-is-what-happens-when-a-drone-slams-into-a-planes-wing-at-speed/
by Rachel Baxter • Originally published October 23, 2018

14) **If You Use Your Web Browser's Incognito Mode, We've Got Bad News**
https://www.iflscience.com/technology/if-you-use-your-web-browsers-incognito-mode-weve-got-bad-news-/
by Aliyah Kovner • Originally published April 26, 2018

15) **Amazon's Alexa Told a Customer to Kill His Foster Parents. Er, What?**
https://www.iflscience.com/space/a-dark-matter-hurricane-appears-to-be-blowing-past-earth-right-now/
by James Felton • Originally published November 13, 2018

16) **Here's Why You Should Probably Wrap Your Car Keys in Tinfoil**
https://www.iflscience.com/technology/heres-why-you-should-probably-wrap-your-car-keys-in-tin-foil/
by Jonathan O'Callaghan • Originally published July 12, 2018

17) **New Fluoride Battery Could Be Charged Just Once a Week**
https://www.iflscience.com/technology/new-fluoride-battery-could-be-charged-just-once-a-week/
by Alfredo Carpineti • Originally published December 7, 2018

APPENDIX

18) **Why Are Mars's Sunsets Blue?**
https://www.iflscience.com/space/why-are-mars-sunsets-blue/
by Alfredo Carpineti • Originally published December 4, 2018

19) **Colossal Drawing of a Penis That Can Be Seen from Space Proves Humanity Will Never Change**
https://www.iflscience.com/space/colossal-drawing-of-a-penis-that-can-be-seen-from-space-proves-humanity-will-never-change/
by Robin Andrews • Originally published June 29, 2018

20) **Astronomers Have Found Another Puzzling "Alien Megastructure" Star**
https://www.iflscience.com/space/astronomers-have-found-another-puzzling-alien-megastructure-star/
by Alfredo Carpineti • Originally published November 22, 2018

21) **Elon Musk's Tesla Roadster Could Crash Back into Earth**
https://www.iflscience.com//space/theres-a-small-chance-elon-musks-tesla-roadster-could-hit-earth/
by Jonathan O'Callaghan • Originally published February 19, 2018

22) ***Voyager 2* Has Just Entered Interstellar Space, NASA Confirms**
https://www.iflscience.com/space/voyager-2-has-just-entered-interstellar-space-nasa-confirms/
by Tom Hale • Originally published December 10, 2018

23) **Uranus Has Experienced a Colossal Pounding**
https://www.iflscience.com/space/uranus-experienced-a-colossal-impact-that-knocked-it-on-its-side/
by Alfredo Carpineti • Originally published July 3, 2018

24) **First Results from NASA's Twins Experiment Surprise Scientists**
https://www.iflscience.com/space/first-results-from-nasas-twins-experiment-surprise-scientists/
by Jonathan O'Callaghan • Originally published January 31, 2017

25) **A Super-Earth Has Been Discovered Just Six Light-Years Away, the Second Closest Planet to Our Solar System**
https://www.iflscience.com/space/a-superearth-has-been-discovered-just-6-lightyears-away-the-second-closest-planet-to-our-solar-system/
by Jonathan O'Callaghan • Originally published November 14, 2018

26) **A Huge Lake of Liquid Water Has Been Found on Mars**
https://www.iflscience.com/space/weve-finally-found-actual-liquid-water-on-mars/
by Jonathan O'Callaghan • Originally published July 25, 2018

27) **A NASA Spacecraft May Have Detected a Giant Wall at the Edge of the Solar System**
https://www.iflscience.com/space/a-nasa-spacecraft-may-have-detected-a-giant-wall-at-the-edge-of-the-solar-system/
by Jonathan O'Callaghan • Originally published August 15, 2018

APPENDIX

28) **Earth Is Passing through a Dark Matter "Hurricane" Right Now**
https://www.iflscience.com/space/a-dark-matter-hurricane-appears-to-be-blowing-past-earth-right-now/
by Jonathan O'Callaghan • Originally published November 13, 2018

29) **A Physicist Claims He's Figured Out Why We Haven't Met Aliens Yet, and It's Pretty Worrying**
https://www.iflscience.com/space/physicist-has-a-new-solution-for-the-fermi-paradox-and-its-pretty-worrying/
by Alfredo Carpineti • Originally published May 29, 2018

30) **Study Reveals That Uranus Smells of Farts**
https://www.iflscience.com/space/study-reveals-uranus-smells-of-farts/
by Alfredo Carpineti • Originally published April 23, 2018

31) **Biblical City of Sin Destroyed by "Sulfur and Fire" May Have Been Flattened by Asteroid**
https://www.iflscience.com/editors-blog/biblical-city-of-sin-destroyed-by-sulfur-and-fire-may-have-been-flattened-by-asteroid-/
by Madison Dapcevich • Originally published November 23, 2018

32) **Astronomers Have Spotted a Mysterious "Ghost" Galaxy Next to the Milky Way**
https://www.iflscience.com/space/a-new-ghostly-galaxy-has-been-spotted-around-the-milky-way/
by Alfredo Carpineti • Originally published November 14, 2018

33) **Declassified Military Report Reveals Extreme Solar Storm Likely Detonated Mines during Vietnam War**
https://www.iflscience.com/editors-blog/declassified-military-report-reveals-extreme-solar-storm-likely-detonated-mines-during-vietnam-war/
by Madison Dapcevich • Originally published November 9, 2018

34) **Somebody Literally Coughed Up a Lung**
https://www.iflscience.com/health-and-medicine/somebody-literally-coughed-up-a-lung/
by Katie Spalding • Originally published December 6, 2018

35) **This Is What's Actually Happening When a Woman "Squirts" During Sex**
https://www.iflscience.com/health-and-medicine/women-squirting-during-sex-may-actually-be-peeing/
by Janet Fang • Originally published January 9, 2015

36) **Genetic Analysis Finally Solves the Mystery of the "Atacama Alien"**
https://www.iflscience.com/health-and-medicine/scientists-have-analyzed-the-genome-of-this-freaky-6inch-skeleton-known-as-the-atacama-alien/
by Aliyah Kovner • Originally published March 22, 2018

37) **How Long Does It Take to Poop Lego?**
https://www.iflscience.com/health-and-medicine/how-long-does-it-take-to-poop-lego/
by Rosie McCall • Originally published November 26, 2018

APPENDIX

38) **Doctor Issues Warning over Dangerous and Deadly Masturbation—But Don't Worry, It's Safe If You Don't Do This!**
https://www.iflscience.com/health-and-medicine/dangerous-masturbation-kills-a-suprising-number-of-people/
by Tom Hale • Originally published February 9, 2018

39) **A Man Took Waaaaaaay Too Much Viagra. Here's What Happened to Him**
https://www.iflscience.com/health-and-medicine/man-takes-too-much-viagra-its-done-something-really-strange-to-his-vision/
by Rosie McCall • Originally published October 2, 2018

40) **One Joint May Be All It Takes to Change the Structure of the Teenage Brain**
https://www.iflscience.com/health-and-medicine/one-joint-may-be-all-it-takes-to-change-the-structure-of-the-teenage-brain/
by Rosie McCall • Originally published January 15, 2019

41) **Apparently Penis Whitening Is a Thing**
https://www.iflscience.com/health-and-medicine/apparently-penis-whitening-is-now-a-thing-in-thailand/
by Rosie McCall • Originally published July 5, 2018

42) **Woman Who Received Lung Transplant Developed Peanut Allergy from Her Donor**
https://www.iflscience.com/health-and-medicine/this-womans-lung-transplant-gave-her-a-peanut-allergy/
by Katie Spalding • Originally published January 2, 2019

43) **Weed or Booze? Scientists Finally Settle Which Is Worse for Your Brain**
https://www.iflscience.com/health-and-medicine/weed-or-booze-scientists-finally-settle-which-is-worse-for-your-brain/
by Madison Dapcevich • Originally published February 20, 2018

44) **Study Finds Spanked Children Are More Likely to Have Developmental Delays**
https://www.iflscience.com/health-and-medicine/study-finds-spanked-children-are-more-likely-to-have-developmental-delays/
by Robin Andrews • Originally published January 23, 2018

45) **Woman Develops Rare Condition That Leaves Her Unable to Hear Men**
https://www.iflscience.com/health-and-medicine/women-develops-rare-condition-that-leaves-her-unable-to-hear-men/
by Rosie McCall • Originally published January 11, 2019

46) **There's Something You Need to Know about the McDonald's Touchscreens**
https://www.iflscience.com/health-and-medicine/theres-something-you-need-to-know-about-the-new-mcdonalds-touchscreens/
by Tom Hale • Originally published November 29, 2018

APPENDIX

47) **A Cancer "Kill Switch" Has Been Found in the Body—And Researchers Are Already Hard at Work to Harness It**
https://www.iflscience.com/health-and-medicine/a-cancer-kill-switch-has-been-found-in-the-body-and-researchers-are-already-hard-at-work-to-harness-it/
by Aliyah Kovner • Originally published November 6, 2018

48) **First-Ever Baby Born Following a Uterus Transplant from a Deceased Donor**
https://www.iflscience.com/health-and-medicine/firstever-baby-born-following-a-uterus-transplant-from-a-deceased-donor-/
by Tom Hale • Originally published December 4, 2018

49) **Hand Dryers Spread Bacteria So Dramatically That Scientists Think They're a Public Health Threat**
https://www.iflscience.com/health-and-medicine/hand-dryers-spread-bacteria-so-dramatically-that-scientists-think-theyre-a-public-health-threat/
by Aliyah Kovner • Originally published September 10, 2018

50) **Brutal Chimpanzee War Was Likely Driven by Power, Ambition, and Jealousy**
https://www.iflscience.com/plants-and-animals/brutal-chimpanzee-war-was-likely-driven-by-power-ambition-and-jealousy
by Josh Davis • Originally published March 27, 2018

51) **Evolution Could Destroy Our Ability to Tolerate Alcohol**
https://www.iflscience.com/plants-and-animals/evolution-could-destroy-our-ability-to-tolerate-alcohol/
by Rosie McCall • Originally published February 20, 2018

52) **Don't Think Arachnids Are Loving? This Spider Nurses Its Young with Milky Liquid**
https://www.iflscience.com/plants-and-animals/dont-think-arachnids-are-loving-this-spider-nurses-its-young-with-milky-liquid/
by Madison Dapcevich • Originally published November 29, 2018

53) **A Scientist Has Been Eaten Alive by a Crocodile**
https://www.iflscience.com/plants-and-animals/a-scientist-has-been-eaten-alive-by-a-crocodile/
by Tom Hale • Originally published January 14, 2019

54) **GM Crops Found to Increase Yields and Reduce Harmful Toxins in 21 Years of Data**
https://www.iflscience.com/plants-and-animals/gm-crops-found-to-increase-yields-and-reduce-harmful-toxins-in-21-years-of-data/
by Jonathan O'Callaghan • Originally published February 21, 2018

55) **The "Reverse Zombie" Tick Is Spreading around America, Causing a Strange Condition As It Goes**
https://www.iflscience.com/plants-and-animals/the-reverse-zombie-tick-is-spreading-around-america-causing-a-very-strange-condition-as-it-goes/
by James Felton • Originally published June 20, 2017

56) **This Is What Eating People Does to the Human Body**
https://www.iflscience.com/plants-and-animals/this-is-what-eating-people-does-to-the-human-body/
by Tom Hale • Originally published March 9, 2018

57) **Mountain Gorillas Are No Longer "Critically Endangered" after a Successful Conservation Effort**
https://www.iflscience.com/plants-and-animals/mountain-gorillas-are-no-longer-critically-endangered-after-a-successful-conservation-effort/
by Jonathan O'Callaghan • Originally published November 14, 2018

58) **Zoo Creates World's First Reptile Swim-Gym to Fight Snake Obesity**
https://www.iflscience.com/plants-and-animals/zoo-creates-worlds-first-reptile-swimgym-to-fight-snake-obesity/
by Stephen Luntz • Originally published January 18, 2019

59) **Why Do Men Have Nipples?**
https://www.iflscience.com/plants-and-animals/why-do-men-have-nipples/
by Rosie McCall • November 24, 2018

60) **Whales Became Really Stressed during World War II, Study Shows**
https://www.iflscience.com/plants-and-animals/whales-got-really-stressed-during-world-war-two-study-shows/
by Jonathan O'Callaghan • Originally published November 22, 2018

61) **Majority of Coffee Species Threatened with Extinction**
https://www.iflscience.com/plants-and-animals/majority-of-coffee-species-threatened-with-extinction/
by Rachel Baxter • Originally published January 17, 2019

62) **Cats Are Not Inherently Antisocial Creatures. It's Just You**
https://www.iflscience.com/plants-and-animals/cats-are-not-inherently-antisocial-creatures-its-just-you/
by Rosie McCall • Originally published January 18, 2019

63) **World's Smallest Dinosaur Footprints Found, Measuring Less Than Half an Inch (1 Cm)**
https://www.iflscience.com/plants-and-animals/worlds-smallest-dinosaur-foot-prints-found-measuring-just-1-centimeter-/
by Katy Evans • Originally published November 21, 2018

64) **Very Good Puppy Digs Up 13,000-Year-Old Mammoth Fossil in Its Owner's Backyard**
https://www.iflscience.com/plants-and-animals/very-good-puppy-digs-up-13000yearold-mammoth-fossil-in-its-owners-backyard/
by Aliyah Kovner • Originally published December 4, 2018

APPENDIX

65) **We Now Know How Wombats Produce Their Unique Cubic Poos**
https://www.iflscience.com/plants-and-animals/we-now-know-how-wombats-produce-their-unique-cubic-poos/
by Stephen Luntz • Originally published November 19, 2018

66) **Step Aside Knickers, There's an Even Bigger Cow in Town Called Dozer**
https://www.iflscience.com/plants-and-animals/step-aside-knickers-theres-an-even-bigger-cow-in-town-called-dozer/
by Jonathan O'Callaghan • Originally published November 30, 2018

67) **Man Who Fell into Yellowstone Hot Spring Completely Dissolved within a Day**
https://www.iflscience.com/environment/man-fell-yellowstone-hot-spring-completely-dissolved-day/
by Robin Andrews • Originally published November 17, 2016

68) **Pompeii Skeleton Reveals the "Unluckiest Guy in History"**
https://www.iflscience.com/editors-blog/pompeii-skeleton-of-unluckiest-guy-in-history-is-the-internets-new-favorite-meme/
by James Felton • Originally published May 30, 2018

69) **Photographer Captures Amazing Images of Weirdly Alien "Light Pillars" Floating in the Sky**
https://www.iflscience.com/environment/photographer-captures-amazing-images-of-weirdly-alien-light-pillars-floating-in-the-sky/
by Jonathan O'Callaghan • Originally published October 26, 2018

70) **Antarctica Is Now Melting Six Times Faster Than It Was in 1979**
https://www.iflscience.com/environment/antarctica-is-now-melting-six-times-faster-than-it-was-in-1979/
by Rachel Baxter • Originally published January 15, 2019

71) **When Was the Worst Time to Be Alive in Human History?**
https://www.iflscience.com/environment/what-was-the-worst-time-to-be-alive-in-human-history/
by Tom Hale • Originally published November 16, 2018

72) **Organic Food Is Worse for the Climate Than Non-Organic Food**
https://www.iflscience.com/environment/organic-food-is-worse-for-the-climate-than-nonorganic-food/
by James Felton • Originally published December 14, 2018

73) **Scientists Have Spotted a "Lost Continent" Using Satellites**
https://www.iflscience.com/environment/ancient-lost-continents-discovered-beneath-the-ice-sheets-of-antarctica/
by Tom Hale • Originally published November 8, 2018

APPENDIX

74) **Huge 210-Foot (64-Meter) Fatberg Discovered beneath Quaint English Seaside Town**
https://www.iflscience.com/environment/huge-64meter-fatberg-discovered-beneath-quaint-english-seaside-town/
by Rosie McCall • Originally published January 9, 2019

75) **Microplastics Found in 100 Percent of Sea Turtles Tested**
https://www.iflscience.com/environment/microplastics-found-in-100-percent-of-sea-turtles-tested/
by Tom Hale • Originally published December 6, 2018

76) **Something Living at the Bottom of the Sea Is Absorbing Large Amounts of the CO_2**
https://www.iflscience.com/environment/something-living-at-the-bottom-of-the-sea-is-absorbing-large-amounts-of-the-co2-in-oceans/
by Madison Dapcevich • Originally published November 21, 2018

77) **A Mystery about Easter Island's Statues Might Finally Be Solved**
https://www.iflscience.com/editors-blog/a-mystery-about-easter-islands-statues-might-finally-be-solved/
by Tom Hale • Originally published June 4, 2018

78) **An Island off the Coast of Japan Has Gone Missing**
https://www.iflscience.com/environment/an-island-off-the-coast-of-japan-has-gone-missing-/
by Rosie McCall • Originally published November 2, 2018

79) **The Map You Grew Up with Is a Lie. This Is What the World Really Looks Like**
https://www.iflscience.com/environment/the-map-you-grew-up-with-is-a-lie-this-is-what-the-world-really-looks-like/
by Tom Hale • Originally published October 22, 2018

80) **2016–2018 Have Been the Hottest Years on Record, UN Report Reveals**
https://www.iflscience.com/environment/the-last-4-years-have-been-the-hottest-on-record-un-report-reveals/
by Tom Hale • Originally published November 30, 2018

81) **Mass Grave of Child Human Sacrifice Victims Found in Peru**
https://www.iflscience.com/editors-blog/mass-grave-of-child-human-sacrifice-victims-found-in-peru/
by Jonathan O'Callaghan • Originally published April 26, 2018

82) **Earth's Magnetic Field Is Up to Some Seriously Weird Stuff and No One Knows Why**
https://www.iflscience.com/environment/earths-magnetic-field-is-up-to-some-seriously-weird-stuff-and-no-one-knows-why-/
by Tom Hale • Originally published January 11, 2019

APPENDIX

83) **New Research Suggests Italian Supervolcano Is Filling Up with Magma**
https://www.iflscience.com/environment/new-research-suggests-italian-supervolcano-is-filling-up-with-magma/
by Alfredo Carpineti • Originally published November 15, 2018

84) **We Just Found the Part of the Brain Responsible for Free Will**
https://www.iflscience.com/brain/we-just-found-the-part-of-the-brain-responsible-for-free-will/
by Rosie McCall • Originally published October 3, 2018

85) **Artificial Intelligence Re-creates Images from inside the Human Brain**
https://www.iflscience.com/brain/artificial-intelligence-recreates-images-from-inside-the-human-brain/
by Jonathan O'Callaghan • Originally published January 3, 2018

86) **Here's a US Army Trick for Falling Asleep Anywhere in 120 Seconds**
https://www.iflscience.com/brain/heres-a-us-army-trick-for-falling-asleep-anywhere-in-120-seconds/
by Jonathan O'Callaghan • Originally published September 4, 2018

87) **Growing Up Poor Physically Changes the Structure of a Child's Brain**
https://www.iflscience.com/brain/growing-up-poor-changes-the-structure-of-a-childs-brain-heres-how/
by Madison Dapcevich • Originally published December 27, 2018

88) **People Would Rather Save a Cat Than a Criminal in Worldwide Trolley Problem Study**
https://www.iflscience.com/brain/people-would-rather-save-a-cat-than-a-criminal-in-worldwide-trolley-problem-study/
by Jonathan O'Callaghan • Originally published October 25, 2018

89) **A Technique to Control Your Dreams Has Been Verified for the First Time**
https://www.iflscience.com/brain/a-technique-to-control-your-dreams-has-been-verified-for-the-first-time/
by Stephen Luntz • Originally published October 19, 2017

90) **Here's What Happens to Alcoholics' Brains When They Quit Drinking**
https://www.iflscience.com/brain/what-happens-alcoholics-brains-when-they-quit-drinking/
by Ben Taub • Originally published March 3, 2016

91) **Pink Isn't Real**
https://www.iflscience.com/brain/pink-isnt-real/
by Tom Hale • Originally published December 19, 2018

92) **The Key to a Happy Sex Life Sounds Pretty Unsexy, According to This Study**
https://www.iflscience.com/brain/the-key-to-a-happy-sex-life-sounds-pretty-unsexy-according-to-this-study/
by Tom Hale • Originally published July 31, 2018

APPENDIX

93) **Microdosing Magic Mushrooms Could Spark Creativity and Boost Cognitive Skills**
https://www.iflscience.com/brain/microdosing-magic-mushrooms-could-spark-creativity-and-boost-cognitive-skills/
by Tom Hale • Originally published October 26, 2018

94) **Whether You're a Go-Getter or a Procrastinator Depends on This**
https://www.iflscience.com/brain/whether-youre-a-gogetter-or-a-procrastinator-depends-on-this-/
by Tom Hale • Originally published August 27, 2018

95) **How and Why Orgasm Faces Differ around the World**
https://www.iflscience.com/brain/how-and-why-orgasm-faces-differ-around-the-world/
by Tom Hale • Originally published October 22, 2018

96) **Scientists Can Read Rats' Minds and Predict Where They Will Go Next**
https://www.iflscience.com/brain/scientists-can-read-rats-minds-and-predict-where-they-will-go-next/
by Alfredo Carpineti • Originally published December 6, 2018

97) **You Can Spot a Narcissist from This Facial Feature, According to New Study**
https://www.iflscience.com/brain/you-can-spot-a-narcissist-from-this-facial-feature-according-to-new-study/
by Tom Hale • Originally published June 4, 2018

98) **Why Do You Lose Your Memory When You Get Really Drunk?**
https://www.iflscience.com/brain/why-do-you-lose-your-memory-when-you-get-really-drunk/
by Aliyah Kovner • Originally published January 1, 2019

99) **This Type of Man Gives the Best Orgasms**
https://www.iflscience.com/brain/what-type-of-men-give-the-best-orgasms-according-to-research/
by Dami Olonisakin • Originally published September 21, 2017

100) **These Personality Traits Could Dictate How Often Men Have Sex, Study Claims**
https://www.iflscience.com/brain/these-personality-traits-could-dictate-how-often-men-have-sex-study-claims/
by Jonathan O'Callaghan • Originally published November 23, 2018

101) **China Just Set a New Nuclear Fusion Record By Reaching Temperatures of 180 Million Degrees**
https://www.iflscience.com/physics/china-just-set-a-new-nuclear-fusion-record-by-reaching-temperatures-of-100-million-degrees-/
by Jonathan O'Callaghan • Originally published November 15, 2018

102) **These Scientists Say They've Invented Something That Can Create Water Out of Desert Air**
https://www.iflscience.com/chemistry/these-scientists-say-theyve-invented-something-that-can-create-water-out-of-desert-air/
by Katie Spalding • Originally published December 5, 2018

APPENDIX

103) **World War II Bombing Raids Were Felt Even at the Edge of Space**
https://www.iflscience.com/physics/world-war-two-bombing-raids-were-felt-even-at-the-edge-of-space/
by Alfredo Carpineti • Originally published September 26, 2018

104) **Amateur Scientists Just Proved Einstein Wrong**
https://www.iflscience.com/physics/one-of-einsteins-major-theories-just-got-disproved-by-a-bunch-of-amateurs/
by Alfredo Carpineti • Originally published May 9, 2018

105) **Residents of UK Town Forced to Evacuate after Cleaning Accident Goes Very Wrong**
https://www.iflscience.com/chemistry/residents-of-uk-town-forced-to-evacuate-after-cleaning-accident-goes-very-wrong/
by Rosie McCall • Originally published January 3, 2019

106) **New Form of Lab-Made Gold Is Better and Golder Than Nature's Pathetic Version**
https://www.iflscience.com/chemistry/new-form-of-labmade-gold-is-better-and-golder-than-natures-pathetic-version/all/
by Robin Andrews • Originally published July 5, 2018

107) **Why Does the Sound of Your Own Recorded Voice Bother You So Much?**
https://www.iflscience.com/physics/why-does-the-sound-of-your-own-recorded-voice-bother-you-so-much/
by Tom Hale • Originally published May 25, 2018

108) **Scientists Say They've Created a Strange New State of Matter That Doesn't Play by the Rules**
https://www.iflscience.com/physics/scientists-say-theyve-created-a-strange-new-state-of-matter-that-doesnt-play-by-the-rules/
by Jonathan O'Callaghan • Originally published November 1, 2018

109) **People Secretly Believe That the Eyes Send Out Force-Carrying Beams**
https://www.iflscience.com/physics/people-secretly-believe-the-eyes-send-out-force-carrying-beams/
by Stephen Luntz • Originally published December 21, 2018

110) **Chinese and Russian Scientists Have Been Heating Huge Portions of the Atmosphere (on Purpose)**
https://www.iflscience.com/physics/chinese-and-russian-scientists-have-been-heating-huge-portions-of-the-atmosphere-on-purpose/
by Katie Spalding • Originally published December 18, 2018

111) **Scientists Have Invented a Fourth Type of Chocolate**
https://www.iflscience.com/chemistry/scientists-invented-fourth-type-chocolate/
by Robin Andrews • Originally published September 17, 2017

112) **Study Claims That the Great Pyramid of Giza Could Focus Some Electromagnetic Waves**
https://www.iflscience.com/physics/study-claims-that-the-great-pyramid-of-giza-could-focus-some-electromagnetic-waves/
by Alfredo Carpineti • Originally published July 31, 2018

113) **Why Do You Get More Static Electric Shocks When It's Cold?**
https://www.iflscience.com/physics/the-reason-you-get-more-static-electric-shocks-in-cold-weather/
by Tom Hale • Originally published February 27, 2018

114) **Hurricane Patricia's Lightning Fired a Beam of Antimatter Down to Earth**
https://www.iflscience.com/physics/hurricane-patricias-lightning-fired-a-beam-of-antimatter-down-to-earth/
by Robin Andrews • Originally published May 24, 2018

115) **Quantum Uncertainty Suggests That an Object Can Be Multiple Temperatures at Once**
https://www.iflscience.com/physics/quantum-uncertainty-suggests-that-an-object-can-be-multiple-temperatures-at-once/
by Alfredo Carpineti • Originally published September 18, 2018

116) **CERN Announces Concept Design for Its 62-Mile (100-Km) Future Collider**
https://www.iflscience.com/physics/cern-announces-concept-design-for-its-100kilometer-future-collider/
by Alfredo Carpineti • Originally published January 16, 2019

ART CREDITS

Illustrations by Tom Rourke except for the following art from Getty Images:

Endpages: CSA-Archive / DigitalVision Vectors
Page 4: Jobalou/ DigitalVision Vectors
Page 34: Adobest/ Getty Images Plus
Page 35: credit: Big_Ryan / DigitalVision Vectors
Page 66: ibusca / DigitalVision Vectors
Page 67: miniature / DigitalVision Vectors
Page 96: bauhaus 1000 / DigitalVision Vectors
Page 96: ivan_baranov / Getty Images Plus
Page 122: Dr. Anthony Romilio
Page 128: filo/ DigitalVision Vectors
Page 129: Yulia_Malinovskaya / Getty Images Plus
Page 151: Neil Kaye/Met Office
Page 158: ilbusca/DigitalVision Vectors
Page 159: ArnaPhoto / Getty Images Plus
Page 218: ©CERN
Page 220: grebeshkovmaxim/ Getty Images Plus

PHOTO CREDITS

Page 9: Exit International
Page 20: worldofstock/Getty Images Plus
Page 26: IBM
Page 29: University of Dayton Research Institute
Page 39: NASA/JPL-Caltech/MSSS/ Texas A&M University
Page 42: Google Maps
Page 45: SpaceX
Page 48: Jacob Kegerreis/Durham University
Page 93: Hospital das Clínicas da Faculdade de Medicina da USP
Page 95: Paul McDougall/Getty Images Plus
Page 106: CHEN Zhanqi
Page 110: Joesboy/Getty Images Plus
Page 114: WLDavies/E+
Page 121: Mykola Sosiukin/Getty Images Plus
Page 126: alessia penny/Getty Images Plus
Page 135: Courtesy of Parco Archeologico di Pompei
Page 136: Vincent Brady Photography
Page 146: mammuth/Getty Images Plus
Page 148: Antonello Proietti/Getty Images Plus
Page 195: 35007/E+
Page 202: ALCHEMIST-HP / WIKIMEDIA COMMONS; CC BY-SA 3.0
Page 210: Sjo / E+

INDEX

A

acetaldehyde, 105
adaptive optics, 82–83
addiction, 172–173
ADELE (Airborne Detector for Energetic Lightning Emissions), 216
agency, 163–164
aircraft, 22–23, 28–30
akinetic mutism, 164
alcohol
    cannabis versus, 86–87
    effects of on brain, 172–173
    memory and, 185–186
    tolerance for, 104–105
alcohol dehydrogenase (ADH), 105
Alexa, 31
Alice UV spectrometer, 56
alien limb syndrome, 164
alien megastructure, 43
allergies
    from ticks, 110–111
    in transplant patient, 85
Amazon Alexa devices, 31
AMPHIBIO, 24–25
anesthetics, 68, 69
Anglo Australian Telescope, 62
animals. *See* plants and animals
Antarctic ice loss, 138–139
anterior cingulate cortex, 164
anthropic principle, 59
antibiotics, 68
antimatter, 191, 215–217
antiseptics, 68
Antlia 2 (Ant 2), 62, 64
archaea, 134

artificial intelligence, 13, 165–166
assisted suicide, 9–10
Atacama alien (Ata), 75–77
atmosphere, heating of, 208–210
atomic bombs, lost, 14–16
"Aussie Weiner," 41–43
autoerotic asphyxiation, 80–81

B

Bachpan Bachao Andolan (BBA), 21
bacteria
    carbon dioxide absorption by, 146–147
    from hand dryers, 95
    on touchscreens, 89–90
Baird, John Logie, 7
Barnard's star, 50
Bell, Alexander Graham, 7
Bell test experiment, 199
Bellino (ox), 127
benthic bacteria, 146–147
Benz, Karl, 7
Big Bang, 37
BIG Bell Test, 199
birth after uterus transplant, 93–95
blockchains, 27
bone conduction, 204
brain
    about, 160–162
    alcohol and, 86–87
    of alcoholics, 172–173
    cannabis and, 84, 86–87
    conscientiousness, 175–176
    dream control, 170–171

    falling asleep, 167
    free will, 163–164
    images from, 165–166
    memory loss due to alcohol, 185–186
    microdosing, 177–178
    narcissistic personality traits, 184
    orgasm faces, 180–181
    orgasms, 187
    personality traits, 187
    pink and, 174
    poverty and development of, 168–169
    procrastination and, 179
    rats, reading minds of, 183–184
    trolley problem, 169
breathing underwater, 24–25
"broken arrow" incidents, 14
BSE (mad cow disease), 113–114

C

calcium chloride, 196
Campi Flegrei, 156–157
cancer "kill switch," 91–92
cannabis, 84, 86–87
cannibalism, 112–114
car keys wrapped in tinfoil, 32–33
carbon dioxide, 146–147
carbon dioxide emissions, 130–131
CASE (Cognitive Architecture for Space Exploration), 13
cats, 121, 169
CERN, 218–219
Cerne Abbas Giant, 43
CES 2019, 16

INDEX

Chelyabinsk, Russia, 61
chemistry. *See* physics and chemistry
children
  brain development in, 168–169
  human sacrifice and, 153
  missing, 20–21
Chimp civil war, 99, 101–103
China Seismo-Electromagnetic Satellite (CSES), 209
chlorine gas, 200–202
chocolate, fourth type of, 211
Clarion-Clipperton Fracture Zone (CCFZ), 147
Clarke, Arthur C., 6
claustrum, 161
cleaning accident, 200–202
climate change, 130–131, 139, 153
cochlea, 204
coffee species, 120
coliform bacteria, 89
Compact Linear Collider (CLIC), 219
computer, world's smallest, 26–27
cone-rod dystrophy, 83
conscientiousness, 175–176
consciousness, 160–161
Consumer Electronics Show (CES), 19
convergent thinking, 178
corporal punishment, 88
cortisol, 118–120
cows, large, 127
creationists, 99
Creutzfeldt-Jakob disease (CJD), 113–114
crocodile, scientist-eating, 108–109
cryptographic anchors, 26–27

D

D1 dopamine receptors, 173
dark matter, 64
dark matter "hurricane," 57
Darwin, Charles, 98
Darwin, Erasmus, 116
data, on Facebook, 18–19
deep neural network (DNN), 161, 165–166
deliquescence, 196
dementia, 114
developmental delays, 88
digestive system, passing Lego through, 78–79
dinosaur footprints, 122–123
DISE (Death by Induced Survival gene Elimination), 91–92
divergent thinking, 178
DNA
  of Atacama alien, 77
  manipulation of, 68
  twins experiment and, 49–50
DNA methylation, 50
dopamine, 172–173
Dozer (cow), 127
dream control, 170–171
Dromaeosauriformipes rarus, 123
drones, colliding with plane, 28–30

E

Earth-like planet, 50
Easter Island statues, 148–150
Eastern Roman Empire, decline of, 142
Einstein, Albert, 7, 36–37, 191–192, 198–199
electroaerodynamic-powered plane, 22–23
electromagnetic spectrum, 174
electromagnetic waves, focusing of, 213–214
electroretinogram, 82–83
entanglement, 198–199
Enterococcus faecalis, 90
environment
  about, 130–132
  Antarctic ice loss, 138–139
  Easter Island statues, 148–150
  fatbergs, 143–144
  hottest years recorded, 153
  human sacrifice, 153
  island, missing, 150
  "light pillars," 136–137
  "lost continent," 142
  magnetic field, 154–155
  maps, misleading, 151–152
  microplastics, 145
  organic food, 142
  Pompeii skeleton, 135
  volcanoes, 156–157
  worst time to be alive, 140–142
  Yellowstone hot spring, 133–134
Equal Earth projection map, 152
Esanbe Hanakita Kojima, 150
ethanol, 185–186
Europa, 37–38
European Strategy for Particle Physics, 219
evolution, 98–99
Exit International, 9

INDEX

ExoMars rover, 54
Experimental Advanced Superconducting Tokamak (EAST), 193–194
extrasolar planets (exoplanets), 37, 48
extraterrestrial life, 37–38, 58–59
eyebrows, distinctive, 184
eyes, invisible beams from, 206–207

F

Facebook, 8, 18–19
facial expressions, 180–181
facial recognition software (FRS), 20–21
falling asleep, 167
famine, 141
Faraday bags, 33
fatbergs, 143–144
female ejaculation, 73–74
Fermi, Enrico, 58, 191
Fermi paradox, 58–59
fluid intelligence, changes in, 178
fluoride liquid batteries, 33
Fore people, 112–114
fossil fuels, 131
Found and Retrieved Time (FART) score, 79
Four-Year War of Gombe, 101–103
fraud prevention, 27
free will, 161, 163–164, 191
freedom-of-choice loophole, 199
functional magnetic resonance imaging (fMRI), 161, 165–166
Future Circular Collider (FCC), 192, 218–219

G

Gaia data, 62, 64
galactose-alpha-1,3-galactose (alpha gal), 110–111
Galen, 68
Gall-Peters world map, 152
general anesthetics, 69
genetic medicine, 68, 75–77
genetically modified crops, 109
geomagnetic pulse, 155
geomagnetic reversal, 155
geometric (feature-based) algorithms, 21
germ theory of disease, 68
"ghost" galaxy, 62, 64
"gills," 24–25
global warming, 130–131
gold, lab-made, 202
Gomorrah, 60–61
Goodall, Jane, 99, 101–103
Gravity Field and Steady-State Ocean Circulation Explore (GOCE) satellite, 142
gravity-mapping data, 142
gray matter, 86, 168
Great Pyramid of Giza, 213–214
guillotine, 7

H

HAL 9000, 13
hand dryers, 95
health and medicine
  about, 68–70
  allergies in transplant patient, 85
  Atacama alien (Ata), 75–77
  birth after uterus transplant, 93–95
  cancer "kill switch," 91–92
  cannabis and, 84
  cannabis vs. alcohol, 86–87
  hand dryers, 95
  hearing loss, 88
  Lego, passing through digestive system, 78–79
  lung, coughing up, 71–72
  masturbation, 80–81
  McDonald's touchscreens, 89–90
  penis whitening, 84
  spanking, 88
  squirting during sex, 73–74
  Viagra overdose, 82–83
hearing, 203–204
hearing loss, 88
heliopause, 55–56
heliosheath, 46
heliosphere, 55
Higgs boson, 218, 219
High-frequency Active Auroral Research Program (HAARP), 208–210
hippocampus, 168, 183–184, 185–186
Hippocrates, 68
Hobbes, Thomas, 163
"hot potting," 133–134
hottest years recorded, 153
Hubble, Edwin, 37
human sacrifice, 153
human waste as fuel, 17
Hurricane Patricia, 215–217
hydrochar, 17
hydrogen, wall of, 56
hydrogen sulfide, 59
hypoxia, 10

INDEX

I

IBM, 26–27
ideal effect, theories of, 181
images from brain, 165–166
inertial bone conduction, 204
insulin, 68
Intergovernmental Panel on Climate Change (IPCC), 130
International Union of Pure and Applied Chemistry (IUPAC), 220
internet browsers, private, 30
interstellar space, 46
involuntary urination, 73–74
ion drive, 22–23
ionosphere, 208–210
island, missing, 150
ITER (International Thermonuclear Experimental Reactor), 194

J

Jupiter, 37

K

Kelly, Mark, 49–50
Kelly, Scott, 49–50
Klebsiella, 90
Knickers (cow), 127
kuru, 112–114

L

lactation
  male, 117
  spider, 106–107
Lake Vostok, 53
Large Hadron Collider (LHC), 192, 218
Large Magellanic Cloud (LMC), 62, 64
Late Antique Little Ice Age, 141
"laughing death," 113
Lego, passing through digestive system, 78–79
Leibniz, Gottfried, 36
"light pillars," 136–137
lightning-generated antimatter, 215–217
liquid light, 205
Listeria, 89
lithium-ion batteries, 33
local realism, principle of, 198–199
lone star tick, 110–111
"lost continent," 142
lucid dreaming, 170–171
lung, coughing up, 71–72
lung transplant, 85
Lyman-alpha line, 56

M

machine learning, 161, 165–166
magic mushrooms, 162, 177–178
magnetic field, 154–155
male nipples, 116–117
mammoth fossil, 124
maps, misleading, 151–152
Markle, Meghan, 11–12
Mars, 37–38, 39–40, 44, 51–54
Mars Advanced Radar for Subsurface and Ionosphere Sounding (MARSIS), 52–53
Mars Express orbiter, 52, 54
mass grave, 153
masturbation, 80–81
matter, new state of, 205
Mayer-Rokitansky-Küster-Hauser (MRKH) syndrome, 94
McDonald's touchscreens, 89–90
meat allergy, 110–111
medicine. *See* health and medicine
memory loss due to alcohol, 185–186
Mercator, Gerardus, 152
Mercator projection, 151–152
Mercury, 40
meteor explosions, 60–61
microdosing, 162, 177–178
microplastics, 145
Microraptors, 123
mindreading, 183–184
mine detonations, 65
mnemonic induction of lucid dreams (MILD), 171
Moon, 48
Moon pillars, 137
Mount Nuovo, 157
Mount Vesuvius, 135
mountain gorillas, 114
Musk, Elon, 38, 44, 46

N

nanostructures, 214
narcissists, 184
NASA Human Research Program, 49–50
National Insurance Crime Bureau, 33
National Oceanic and Atmospheric Administration (NOAA), 155, 215–217
natural selection, theory of, 98
Nazca Lines, 43
Netflix, 16
neural plasticity, 186
neurotransmitters, 160

INDEX

New Horizons, 55–56
Newton, Isaac, 36
nipples, male, 116–117
nuclear fusion, 193–194
nuclear weapons, lost, 14–16

O

obesity, snake, 115
obesity epidemic, 99
On the Fabric of the Human Body (Vesalius), 68
opioid epidemic, 87
optimal coherence tomography, 82–83
orbital mechanics, 44, 46
organic food, 142
orgasm faces, 180–181
orgasms, 187

P

painkillers, 68
pair production, 216
panspermia, 98
parbuckling technique, 150
passwords, sharing of, 16
penis
  colossal drawing of, 41–43
  whitening of, 84
periodic table, 220
Phlegraean Fields, 156–157
photometric algorithms, 21
photosynthesis, 131
physics and chemistry
  about, 190–192
  atmosphere, heating of, 210
  chlorine gas, 200–202
  chocolate, fourth type of, 211
  eyes, invisible beams from, 206–207
  Future Circular Collider (FCC), 218–219
  gold, lab-made, 202
  Great Pyramid of Giza, 213–214
  lightning-generated antimatter, 215–217
  liquid light, 205
  nuclear fusion, 193–194
  periodic table, 220
  proving Einstein wrong, 198–199
  static electricity, 214
  uncertainty principle, 217
  voice, sound of recorded, 203–204
  water, creating out of air, 195–196
  WWII bombing raids, 197
pink, 174
place cells, 183–184
plants and animals
  about, 98–100
  alcohol tolerance, 104–105
  cannibalism, 112–114
  cats, 121
  Chimp civil war, 101–103
  coffee species, 120
  cows, large, 127
  crocodile, scientist-eating, 108–109
  dinosaur footprints, 122–123
  genetically modified crops, 109
  male nipples, 116–117
  mammoth fossil, 124
  mountain gorillas, 114
  snake obesity, 115
  spiders, lactation and, 106–107
  ticks, 110–111
  whales, 118–120
  wombats, 125–126
polaritons, 205
Pompeii skeleton, 135
Pornhub, 8, 11–12
positrons, 216
poverty and brain development, 168–169
precuneus cortex, 164
prefrontal cortex, 168
prions, 113–114
procrastination, 179
Promobot, 19
prostatic-specific antigen (PSA), 74
proton colliders, 219
Pseudomonas, 90
pukao, 148–150
pyramidal neurons, 185–186
pyramids, 213–214

Q

qualia, 160, 161
quantum mechanics, 6–7, 161, 191–192, 198–199, 217

R

reading rats' minds, 183–184
reality testing, 171
recurring slope lineae (RSL), 54
reference memory, 183–184
relativity, theory of, 7, 36–37
relay attacks, 32–33
resonance, 213–214
"reverse zombie" ticks, 110–111

INDEX

reverse-slope hearing loss (RSHL), 88
RNA (ribonucleic acid), 91–92
"RNA world" hypothesis, 98
royal wedding, 11–12
"ruby chocolate," 211

S

S1 stream, 57
Sarco, 9–10
Sasanian Empire, collapse of, 142
Saturn, 40
sea levels, rising, 130, 132, 138–139
self-driving cars, 19
self-order touchscreens, 89–90
semiconductors, 6–7
sex
    conscientiousness and, 175–176
    frequency of, 187
    squirting during, 73–74
    type of man and, 187
sildenafil citrate (Viagra), 82–83
siRNAs, 91–92
Skene glands, 74
sleep, 69
snake obesity, 115
socioeconomic status (SES), effects of, 168–169
Sodom, 60–61
solar flares, 65
space
    about, 36–38
    alien megastructure, 43
    dark matter "hurricane," 57
    extraterrestrial life, 58–59
    "ghost" galaxy, 62, 64
    hydrogen "wall," 55–56
    meteor explosions, 60–61
    penis, colossal drawing of, 41–43
    solar flares, 65
    sunsets on Mars, 39–40
    super-Earth, 50
    Tesla Roadster, 44, 46
    twins experiment, 49–50
    Uranus, 47–48, 59
    Voyager 2, leaving solar system, 46
    water on Mars, 51–54
SpaceX, 38, 44
spanking, 88
spiders, lactation and, 106–107
squirting during sex, 73–74
SRY gene, 117
Staphylococcus, 89–90
static electricity, 214
Stony Tunguska River, 61
Stool Hardness and Transit (SHAT) score, 78–79
stress
    memory formation and, 186
    in whales, 118–120
suicide machine, 9–10
Sun pillars, 137
sunsets on Mars, 39–40
super-Earth, 50
superfluids, 205
supervolano, 156–157
Synamedia, 16

T

Tall el-Hamman, 61
technology
    about, 6–8
    AI, 13
    Alexa, 31
    breathing underwater, 24–25
    computer, world's smallest, 26–27
    drone strikes, 28–30
    Facebook data, 18–19
    facial recognition software, 20–21
    fluoride liquid batteries, 33
    human waste as fuel, 17
    ion drive, 22–23
    key fobs, 32–33
    nuclear weapons, lost, 14–16
    password sharing, 16
    Pornhub data, 11–12
    private internet browsers, 30
    self-driving cars, 19
    suicide machine, 9–10
telephone, 7
television, 7
telomeres, 49–50
tennessine, 220
terrestrial gamma-ray flash (TGF), 216–217
Tesla, 19
Tesla Roadster, 44, 46
thermal expansion, 139
ticks, 110–111
tinfoil-wrapped car keys, 32–33
Titan, 40
tokamak reactor, 193–194
touchscreens, McDonald's, 89–90
Toxeus magnus, 106–107
TrackChild, 20–21
transistors, 7
transplant patients, 85, 93–95
trolley problem, 169
turtles, microplastics in, 145
twins experiment, 49–50

INDEX

U

ultraviolet light, 56
uncertainty principle, 217
universal healthcare, 68–69
universe, mathematical models of, 37
University of Dayton Research Institute (UDRI), 28–30
unmanned aerial vehicles (UAVs), 28–30
Unmanned Systems Academic Summit, 28
Uranus, 47–48, 59
urination, involuntary, 73–74
uterus transplant, 93–95

V

vacuum cleaners, masturbation with, 81
Venus, 40, 46
Vesalius, Andreas, 68
Viagra overdose, 82–83
Vietnam War, 65
Virgin Galactic, 38
vision, Viagra overdose and, 82–83
voice, sound of recorded, 203–204
volcanic winter, 141–142
volcanoes, 156–157
volition, 163–164
Voyager 1, 56
Voyager 2, 46, 56

W

"wall" in space, 55–56
water
  creating out of air, 195–196
  on Mars, 51–54
water gym for snakes, 115
weather
  climate versus, 131
  extreme, 141
web browsers, private, 30
whales during WWII, 118–120
Which? 33
white matter, 86
Whitechapel fatberg, 143
wombats, 100, 125–126
working memory, 183–184
World Magnetic Model (WMM), 155
World Meteorological Organization (WMO), 153
World War II
  bombing raids during, 197
  whales and, 118–120
worst time to be alive, 140–142
Yellowstone hot spring, 133–134
Ypres, Second Battle of, 201

[241]

ACKNOWLEDGMENTS

Now is the time for us to recognize those who have worked hard and stayed true to IFLScience. We have to start with the brains of the organization, Elise Andrew. She created the powerhouse that is IFLScience and without her the world of science would just be dull and f*#king boring!

The next important person in the IFLScience family is the great MD, David Dunhill, whose support and guidance over the years has helped mould the brand into what it is today.

Next, to all the staff, both current and former, at IFLS in the UK, USA, and our lone ranger in Australia! A big thank you to everyone for all your hard work and dedication over the years.

Thank you to the team at Running Press: Cindy De La Hoz, Susan Van Horn, Kristin Kiser, Jessica Schmidt, Amy Cianfrone, and Jennifer Leczkowski. We look forward to our continued relationship and bringing more books to our fans! And to our partners in the States, One Entertainment, thank you for your continued hard work.

Lastly, to our fans. Without you we wouldn't be here today! Thank you to every one of you who has followed us over the years and been part of our journey. We hope you have enjoyed our first ever IFLScience book!